SPARK

'Forget fish oil and sudoku - it's exercise that makes you brainier . . . Regular exercise isn't just good for your body . . . it can also dramatically improve your brain, boosting learning capabilities, reducing stress, smoothing hormonal fluctuations and reversing the signs of ageing . . . This book is the first time scientific evidence from all over the world has been pulled together to show that the fitter you are, the better your brain works'   *Daily Mail*

'Accessible and thought-provoking . . . Dr John Ratey . . . describes a new approach to physical education that builds the link between exercise and the brain . . . He shows that walking, running and strength training not only improve school performance, they also fight stress, anxiety, depression, addiction and even the effects of ageing . . . If exercise came in pill form, it would be plastered across the front page, hailed as the blockbuster drug of the century. So what are you waiting for? Get moving!'   *Focus Magazine*

'Want to raise your IQ? Pump up your heart rate! In *Spark*, Dr. Ratey reveals how cardiovascular exercise can help ward off anxiety, dementia and depression, all while increasing your intelligence. Buy it if you need a little bit of extra motivation to hit the gym'   *Self Magazine*

'At a time of year when the prospect of the gym or a bicycle ride seems particularly unappealing just bear in mind that increasing your metabolic rate will do more than simply decrease your waistline, it might literally change the way you think'   *Birmingham Life*

SPARK

THE REVOLUTIONARY NEW SCIENCE OF EXERCISE AND THE BRAIN

JOHN J. RATEY, MD

with ERIC HAGERMAN

Quercus

First published in Great Britain in 2009 by Quercus
This paperback edition first published in 2010 by

Quercus
21 Bloomsbury Square
London
WC1A 2NS

A CIP catalogue record for this book is available
from the British Library

ISBN 978 1 84916 157 2

Printed and bound in Great Britain by Clays Ltd, St Ives plc

10 9 8 7 6 5 4 3 2 1

To Kenneth Cooper, Carl Cotman, and Phil Lawler,
three revolutionaries without whom this book
could not have been written

In order for man to succeed in life, God provided him with two means, education and physical activity. Not separately, one for the soul and the other for the body, but for the two together. With these two means, man can attain perfection.

—*Plato*

Contents

SPARK

Introduction
Making the Connection

WE ALL KNOW that exercise makes us feel better, but most of us have no idea why. We assume it's because we're burning off stress or reducing muscle tension or boosting endorphins, and we leave it at that. But the real reason we feel so good when we get our blood pumping is that it makes the brain function at its best, and in my view, this benefit of physical activity is far more important—and fascinating—than what it does for the body. Building muscles and conditioning the heart and lungs are essentially side effects. I often tell my patients that the point of exercise is to build and condition the brain.

In today's technology-driven, plasma-screened-in world, it's easy to forget that we are born movers—animals, in fact—because we've engineered movement right out of our lives. Ironically, the human capacity to dream and plan and create the very society that shields us from our biological imperative to move is rooted in the areas of the brain that govern movement. As we adapted to an ever-changing environment over the past half million years, our thinking brain evolved from the need to hone motor skills. We envision our hunter-gatherer ancestors as brutes who relied primarily on physical prowess, but to survive over the long haul they had to use their smarts to find and store food. The relationship between food, physical activity, and learning is hardwired into the brain's circuitry.

But we no longer hunt and gather, and that's a problem. The sedentary character of modern life is a disruption of our nature, and it poses one of the biggest threats to our continued survival. Evidence of this is everywhere: 65 percent of our nation's adults are overweight or obese, and 10 percent of the population has type 2 diabetes, a preventable and ruinous disease that stems from inactivity and poor nutrition. Once an affliction almost exclusively of the middle-aged, it's now becoming an epidemic among children. We're literally killing ourselves, and it's a problem throughout the developed world, not merely a province of the supersize lifestyle in the United States. What's even more disturbing, and what virtually no one recognizes, is that inactivity is killing our brains too — physically shriveling them.

Our culture treats the mind and body as if they are separate entities, and I want to reconnect the two. The mind-body connection has fascinated me for years. My very first lecture, to fellow medical professionals at Harvard, in 1984, was titled "The Body and Psychiatry." It focused on a novel drug treatment, for aggression, that affected both the body and the brain, which I stumbled on as a resident working in the Massachusetts state hospital system. My experience working with the most complicated psychiatric patients set me on a path of investigation into the ways in which treating the body can transform the mind. It's been an enthralling journey, and though it continues, it's time to deliver that message to the public. What neuroscientists have discovered in the past five years alone paints a riveting picture of the biological relationship between the body, the brain, and the mind.

To keep our brains at peak performance, our bodies need to work hard. In *Spark*, I'll demonstrate how and why physical activity is crucial to the way we think and feel. I'll explain the science of how exercise cues the building blocks of learning in the brain; how it affects mood, anxiety, and attention; how it guards against stress and reverses some of the effects of aging in the brain; and how in women it can help stave off the sometimes tumultuous effects of hormonal changes. I'm not talking about the fuzzy notion of runner's high. I'm not talking about a notion at all. These are tangible changes, measured in lab rats and identified in people.

It was already known that exercise increases levels of serotonin, norepinephrine, and dopamine—important neurotransmitters that traffic in thoughts and emotions. You've probably heard of serotonin, and maybe you know that a lack of it is associated with depression, but even many psychiatrists I meet don't know the rest. They don't know that toxic levels of stress erode the connections between the billions of nerve cells in the brain or that chronic depression shrinks certain areas of the brain. And they don't know that, conversely, exercise unleashes a cascade of neurochemicals and growth factors that can reverse this process, physically bolstering the brain's infrastructure. In fact, the brain responds like muscles do, growing with use, withering with inactivity. The neurons in the brain connect to one another through "leaves" on treelike branches, and exercise causes those branches to grow and bloom with new buds, thus enhancing brain function at a fundamental level.

Neuroscientists have just begun studying exercise's impact *within* brain cells—at the genes themselves. Even there, in the roots of our biology, they've found signs of the body's influence on the mind. It turns out that moving our muscles produces proteins that travel through the bloodstream and into the brain, where they play pivotal roles in the mechanisms of our highest thought processes. They bear names such as insulin-like growth factor (IGF-1) and vascular endothelial growth factor (VEGF), and they provide an unprecedented view of the mind-body connection. It's only in the past few years that neuroscientists have begun to describe these factors and how they work, and each new discovery adds awe-inspiring depth to the picture. There's still much we don't understand about what happens in the microenvironment of the brain, but I think what we do know can change people's lives. And maybe society itself.

Why should you care about how your brain works? For one thing, it's running the show. Right now the front of your brain is firing signals about what you're reading, and how much of it you soak up has a lot to do with whether there is a proper balance of neurochemicals and growth factors to bind neurons together. Exercise has a documented, dramatic effect on these essential ingredients. It sets the stage, and when you sit down to learn something new, that stimulation strengthens the relevant connections; with practice, the circuit develops definition, as if you're wearing down a path through a forest. The importance of making these connections carries over to all of the issues I deal with in this book. In order to cope with anxiousness, for instance, you need to let certain well-worn paths grow over while you blaze alternate trails. By understanding such interactions between your body and your brain, you can manage the process, handle problems, and get your mind humming along smoothly. If you had half an hour of exercise this morning, you're in the right frame of mind to sit still and focus on this paragraph, and your brain is far more equipped to remember it.

Everything I have written over the past fifteen years has been aimed at educating people about their brains. Your life changes when you have a working knowledge of your brain. It takes guilt out of the equation when you recognize that there's a biological basis for certain emotional issues. On the other hand, you won't be left feeling helpless when you see how you can influence that biology. This is a point that I keep coming back to with my patients, because people tend to picture the brain as a commander mysteriously issuing orders from an ivory tower, untouchable from the outside. Not at all. Exercise breaks down those barriers. My hope is that if you understand how physical activity improves brain function, you'll be motivated to include it in your life in a positive way, rather than think of it as something you *should* do. Of course you should exercise, but I won't be preaching here. (It probably wouldn't help: experiments with lab rats suggest that forced exercise doesn't do the trick quite like voluntary exercise.) If you can get to the point where you're consistently saying to yourself exercise is something you *want* to do, then you're charting a course to a different future—one that's less about surviving and more about thriving.

In October of 2000 researchers from Duke University made the *New York Times* with a study showing that exercise is better than sertraline (Zoloft) at treating depression. What great news! Unfortunately, it was buried on page fourteen of the Health and Fitness section. If exercise came in pill form, it would be plastered across the front page, hailed as the blockbuster drug of the century.

Other fragments of the story I'm presenting bubble to the surface, only to sink back down. *ABC World News* reports that exercise might stave off Alzheimer's disease in rats; CNN flashes stats on the ever-expanding obesity crisis; the *New York Times* investigates the practice of treating bipolar kids with costly drugs that are only marginally effective yet carry horrendous side effects. What gets lost is that these seemingly unrelated threads are tied together at a fundamental level of biology. I'll explain how, by exploring volumes of new research that hasn't yet appeared anywhere for the general public.

What I aim to do here is to deliver in plain English the inspiring science connecting exercise and the brain and to demonstrate how it plays out in the lives of real people. I want to cement the idea that exercise has a profound impact on cognitive abilities and mental health. It is simply one of the best treatments we have for most psychiatric problems.

I've witnessed this among my patients and my friends, a number of whom have given me permission to tell their stories here. Yet it was far beyond the walls of my office that I discovered the exemplar case study, in a suburban school district outside Chicago. The implications of the most exciting new research merge in this tale of a revolutionary physical education program. In Naperville, Illinois, gym class has transformed the student body of nineteen thousand into perhaps the fittest in the nation. Among one entire class of sophomores, only 3 percent were overweight, versus the national average of 30 percent. What's more surprising—stunning—is that the program has also turned those students into some of the smartest in the nation. In 1999 Naperville's eighth graders were among some 230,000 students from around the world who took an international standards test called TIMSS (Trends in International Mathematics and Science Study), which evaluates knowledge of math and science. In recent years, students in China, Japan, and Singapore have outpaced American kids in these crucial subjects, but Naperville is the conspicuous exception: when its students took the TIMSS, they finished sixth in math and first in the world in science. As politicians and pundits sound the alarm about faltering education in the United States, and about our students being ill-equipped to succeed in today's technology-driven economy, Naperville stands out as an extraordinary bit of good news.

I haven't seen anything as uplifting and inspiring as Naperville's program in decades. At a time when we're bombarded with sad news about overweight, unmotivated, and underachieving adolescents, this example offers real hope. In the first chapter, I'll take you to Naperville. It is the spark that inspired me to write this book.

1
Welcome to the Revolution
A Case Study on Exercise and the Brain

ON A SLIGHT swell of land west of Chicago stands a brick build-ing, Naperville Central High School, which harbors in its base-ment a low-ceilinged, windowless room crowded with treadmills and stationary bikes. The old cafeteria—its capacity long dwarfed by enrollment numbers—now serves as the school's "cardio room." It is 7:10 a.m., and for the small band of newly minted freshmen lounging half asleep on the exercise equipment, that means it's time for gym.

A trim young physical education teacher named Neil Duncan lays out the morning's assignment: "OK, once you're done with your warm-up, we're going to head out to the track and run the *mile*," he says, presenting a black satchel full of chest straps and digital watches—heart rate monitors of the type used by avid ath-letes to gauge their physical exertion. "Every time you go around the track, hit the *red* button. What that's going to *do*—it's going to give you a *split*. It's going to tell you, this is how fast I did my first lap, second lap, third lap. On the fourth and final lap—which will be just as fast if you do it right—" he says, pausing to survey his sleepy charges, "you hit the *blue* button, OK? And that'll stop your watch. Your goal is—well, to try to run your fastest mile. Last but not least, your average heart rate should be above 185."

Filing past Mr. Duncan, the freshmen lumber upstairs, push through a set of heavy metal doors, and in scattered groups they hit the track under the mottled skies of a crisp October morning. Perfect conditions for a revolution.

This is not good old gym class. This is Zero Hour PE, the latest in a long line of educational experiments conducted by a group of maverick physical education teachers who have turned the nineteen thousand students in Naperville District 203 into the fittest in the nation—and also some of the smartest. (The name of the class refers to its scheduled time before first period.) The objective of Zero Hour is to determine whether working out before school gives these kids a boost in reading ability and in the rest of their subjects.

The notion that it might is supported by emerging research showing that physical activity sparks biological changes that encourage brain cells to bind to one another. For the brain to learn, these connections must be made; they reflect the brain's fundamental ability to adapt to challenges. The more neuroscientists discover about this process, the clearer it becomes that exercise provides an unparalleled stimulus, creating an environment in which the brain is ready, willing, and able to learn. Aerobic activity has a dramatic effect on adaptation, regulating systems that might be out of balance and optimizing those that are not—it's an indispensable tool for anyone who wants to reach his or her full potential.

Out at the track, the freckled and bespectacled Mr. Duncan supervises as his students run their laps.

"My watch isn't reading," says one of the boys as he jogs past.

"*Red* button," shouts Duncan. "Hit the *red* button! At the end, hit the *blue* button."

Two girls named Michelle and Krissy pass by, shuffling along side by side.

A kid with unlaced skateboarding shoes finishes his laps and turns in his watch. His time reads eight minutes, thirty seconds.

Next comes a husky boy in baggy shorts.

"Bring it on in, Doug," Duncan says. "What'd you get?"

"Nine minutes."

"Flat?"

"Yeah."

"Nice work."

When Michelle and Krissy finally saunter over, Duncan asks for their times, but Michelle's watch is still running. Apparently, she didn't hit the *blue* button. Krissy did, though, and their times are the same. She holds up her wrist for Duncan. "Ten twelve," he says, noting the time on his clipboard. What he doesn't say is "It looked like you two were really loafing around out there!"

The fact is, they weren't. When Duncan downloads Michelle's monitor, he'll find that her average heart rate during her ten-minute mile was 191, a serious workout for even a trained athlete. She gets an A for the day.

The kids in Zero Hour, hearty volunteers from a group of freshmen required to take a literacy class to bring their reading comprehension up to par, work out at a higher intensity than Central's other PE students. They're required to stay between 80 and 90 percent of their maximum heart rate. "What we're really doing is trying to get them prepared to learn, through rigorous exercise," says Duncan. "Basically, we're getting them to that state of heightened awareness and then sending them off to class."

How do they feel about being Mr. Duncan's guinea pigs? "I guess it's OK," says Michelle. "Besides getting up early and being all sweaty and gross, I'm more awake during the day. I mean, I was cranky all the time last year."

Beyond improving her mood, it will turn out, Michelle is also doing much better with her reading. And so are her Zero Hour classmates: at the end of the semester, they'll show a 17 percent improvement in reading and comprehension, compared with a 10.7 percent improvement among the other literacy students who opted to sleep in and take standard phys ed.

The administration is so impressed that it incorporates Zero Hour into the high school curriculum as a first-period literacy class called Learning Readiness PE. And the experiment continues. The literacy students are split into two classes: one second period, when they're still feeling the effects of the exercise, and one eighth period. As expected, the second-period literacy class performs best. The strategy spreads beyond freshmen who need to boost their reading scores, and guidance counselors begin suggesting that all students schedule their hardest subjects immediately after gym, to capitalize on the beneficial effects of exercise.

It's a truly revolutionary concept from which we can all learn.

FIRST-CLASS PERFORMANCE

Zero Hour grew out of Naperville District 203's unique approach to physical education, which has gained national attention and become the model for a type of gym class that I suspect would be unrecognizable to any adult reading this. No getting nailed in dodgeball, no flunking for not showering, no living in fear of being the last kid picked.

The essence of physical education in Naperville 203 is teaching fitness instead of sports. The underlying philosophy is that if physical education class can be used to instruct kids how to monitor and maintain their own health and fitness, then the lessons they learn will serve them for life. And probably a longer and happier life at that. What's being taught, really, is a lifestyle. The students are developing healthy habits, skills, and a sense of fun, along with a knowledge of how their bodies work. Naperville's gym teachers are opening up new vistas for their students by exposing them to such a wide range of activities that they can't help but find something they enjoy. They're getting kids hooked on moving instead of sitting in front of the television. This couldn't be more important, particularly since statistics show that children who exercise regularly are likely to do the same as adults.

But it's the impact of the fitness-based approach on the kids while they're still in school that initially grabbed my attention. The New PE curriculum has been in place for seventeen years now, and its effects have shown up in some unexpected places — namely, the classroom.

It's no coincidence that, academically, the district consistently ranks among the state's top ten, even though the amount of money it spends on each pupil — considered by educators to be a clear predictor of success — is notably lower than other top-tier Illinois public schools. Naperville 203 includes fourteen elementary schools, five junior highs, and two high schools. For the sake of comparison, let's look at Naperville Central High School, where Zero Hour began. Its per-pupil operating expense in 2005 was $8,939 versus $15,403 at Evanston's New Trier High School. New Trier kids scored on average two points higher on their ACT college entrance exams (26.8), but they fared worse than Central's kids on a composite of mandatory state tests, which are taken by every student, not just those applying to college. And Central's composite ACT score for the graduating class of 2005 was 24.8, well above the state average of 20.1.

Those exams aren't nearly as telling as the Trends in International Mathematics and Science Study (TIMSS), a test designed to compare students' knowledge levels from different countries in two key subject areas. This is the exam cited by *New York Times* editorialist Thomas Friedman, author of *The World Is Flat*, when he laments that students in places like Singapore are "eating our lunch." The education gap between the United States and Asia is widening, Friedman points out. Whereas in some Asian countries nearly half of the students score in the top tier, only 7 percent of U.S. students hit that mark.

TIMSS has been administered every four years since 1995. The 1999 edition included 230,000 students from thirty-eight countries, 59,000 of whom were from the United States. While New Trier and eighteen other schools along Chicago's wealthy North Shore formed a consortium to take the TIMSS (thereby masking individual schools' performance), Naperville 203 signed up on its own to get an international benchmark of its students' performance. Some 97 percent of its eighth graders took the test—not merely the best and the brightest. How did they stack up? On the science section of the TIMSS, Naperville's students finished first, just ahead of Singapore, and then the North Shore consortium. *Number one in the world.* On the math section, Naperville scored sixth, behind only Singapore, Korea, Taiwan, Hong Kong, and Japan.

As a whole, U.S. students ranked eighteenth in science and nineteenth in math, with districts from Jersey City and Miami scoring dead last in science and math, respectively. "We have huge discrepancies among our school districts in the United States," says Ina Mullis, who is a codirector of TIMSS. "It's a good thing that we've at least got some Napervilles—it shows that it can be done."

I won't go so far as to say that Naperville's kids are brilliant specifically because they participate in an unusual physical education program. There are many factors that inform academic achievement. To be sure, Naperville 203 is a demographically advantaged school district: 83 percent white, with only 2.6 percent in the low income range, compared with 40 percent in that range for Illinois as a whole. Its two high schools boast a 97 percent graduation rate. And the town's major employers are science-centric companies such as Argonne, Fermilab, and Lucent Technologies, which suggests that the parents of many Naperville kids are highly educated. The deck—in terms of both environment and genetics—is stacked in Naperville's favor.

On the other hand, when we look at Naperville, two factors really stand out: its unusual brand of physical education and its test scores. The correlation is simply too intriguing to dismiss, and I couldn't resist visiting Naperville to see for myself what was happening there. I've long been aware of the TIMSS test and how it points to the failings of public education in this country. Yet the Naperville 203 kids aced the test. Why? It's not as if Naperville is the only wealthy suburb in the country with intelligent, educated parents. And in poor districts where Naperville-style PE has taken root, such as Titusville, Pennsylvania (which I'll discuss later), test scores have improved measurably. My conviction, and my attraction to Naperville, is that its focus on fitness plays a pivotal role in its students' academic achievements.

THE NEW PE

The Naperville revolution started, as such things often do, with equal parts idealism and self-preservation. A visionary junior high physical education teacher named Phil Lawler got the movement off the ground after he came across a newspaper article in 1990 reporting that the health of U.S. children was declining.

"It said the reason they weren't healthy was that they weren't very active," recalls Lawler, a tall man in his fifties, with rimless glasses, who dresses in khakis and white sneakers. "These days everybody knows we have an obesity epidemic," he continues. "But pick up a paper seventeen years ago and that kind of article was unusual. We said, We have these kids every day; shouldn't we be able to affect their health? If this is our business, I thought, we're going bankrupt."

He already felt like his profession received no respect; schools had started cutting phys ed from the curriculum, and now this. A former college baseball pitcher who missed out on the majors, Lawler is a sincere salesman and a natural leader who became a gym teacher to stay close to sports. In addition to teaching PE at District 203's Madison Junior High, he coached Naperville Central's baseball team and served as the district coordinator for PE, but even in these respectable posts, sometimes he was embarrassed to admit what he did for a living. Part of what he saw in that article was an opportunity—a chance to make his job matter.

When Lawler and his staff at Madison took a close look at what was happening in gym, they saw a lot of inactivity. It's the nature of team sports: waiting for a turn at bat, waiting for the center's snap, waiting for the soccer ball to come your way. Most of the time, most of the players just stood around. So Lawler decided to shift the focus to cardiovascular fitness, and he instituted a radical new feature to the curriculum. Once a week in gym class, the kids would run the mile. Every single week! His decision met with groans from students, complaints from parents, and notes from doctors.

He was undeterred, yet he quickly recognized that the grading scale discouraged the slowest runners. To offer nonathletes a shot at good marks, the department bought a couple of Schwinn Airdyne bikes and allowed students to earn extra credit. They could come in on their own time and ride five miles to raise their grades. "So any kid who wanted to get an A could get an A if he worked for it," Lawler explains. "Somewhere in this process, we got into personal bests. Anytime you got a personal best, no matter what it was, you moved up a letter grade." And this led to the founding principle of the approach he dubbed the New PE: Students would be assessed on effort rather than skill. You didn't have to be a natural athlete to do well in gym.

But how does one judge the individual effort of forty kids at a time? Lawler found his answer at a physical education conference he organized every spring. He worked hard to turn the event into an exchange of fresh ideas and technologies, and to encourage attendance he talked the vendors into donating door prizes. Each year at the beginning of the conference, he would push a towel cart through the aisles, collecting bats and balls and other sporting goods. Cast in among the bounty one year was a newfangled heart rate monitor, which at the time was worth hundreds of dollars. He couldn't help himself; he stole it for the revolution. "I saw that son of a buck," he freely admits, "and I said, That's a door prize for Madison Junior High!"

During the weekly mile, he tested the device on a sixth-grade girl who was thin but not the least bit athletic. When Lawler downloaded her stats, he couldn't believe what he found. "Her average heart rate was 187!" he exclaims. As an eleven-year-old, her maximum heart rate would have been roughly 209, meaning she was plugging away pretty close to full tilt. "When she crossed the finish line, she went up to 207," Lawler continues. "*Ding, ding, ding!* I said, You gotta be *kidding* me! Normally, I would have gone to that girl and said, You need to get your ass in gear, little lady! It was really that moment that caused dramatic changes in our overall program. The heart rate monitors were a springboard for everything. I started thinking back to all the kids we must have turned off to exercise because we weren't able to give them credit. I didn't have an athlete in class who knew how to work as hard as that little girl."

He realized that being fast didn't necessarily have anything to do with being fit.

One of Lawler's favorite statistics is that less than 3 percent of adults over the age of twenty-four stay in shape through playing team sports, and this underscores the failings of traditional gym class. But he knew he couldn't have the students run the mile every day, so he set up a program of what they have termed "small-sided sports" — three-on-three basketball or four-on-four soccer — where the students are constantly moving. "We still play sports," Lawler says. "We just do them within a fitness model." Instead of being tested on such trivia as the dimensions of a regulation volleyball court, Naperville's gym students are graded on how much time they spend in their target heart rate zones during any given activity.

"We developed the program not knowing what we were doing," Lawler says. And yet, the New PE has managed to put into practice principles consistent with all the new research about exercise and the brain.

CARRYING THE TORCH

Every revolutionary leader needs a lieutenant, and Lawler couldn't have chosen a more able agitator than Paul Zientarski, Naperville Central High School's physical education coordinator and former football coach. To students and colleagues, Zientarski is Mr. Z, a gray-haired furnace of a man with steady eyes and a facts-is-facts delivery. He has the presence of Mike Ditka and Bill Parcells rolled into one formidable figure of authority. "It took me the longest time to convince him of this stuff," says Lawler of his friend and ally. "But once he buys into it, get out of his way. Because he's going to shove it down your throat if he has to."

As their movement grew, Lawler would take the lead in proselytizing the outside world with the fitness-not-sports message, talking to *Newsweek* and testifying before the U.S. Senate, and Zientarski would become the unwavering enforcer of the mission back home, transforming the phys ed program at Naperville Central into a well-oiled working model of the New PE. Lawler retired from teaching in 2004 after being diagnosed with colon cancer, but he has continued to lobby for daily physical education even during his back-and-forth battle with the disease.

They've both become grassroots experts on the subject of exercise and the brain. They learned by grilling speakers from the conferences Lawler organized, attending sports physiology seminars, reading neuroscience research papers, and constantly e-mailing their findings to each other. And they've taken it upon themselves to educate their colleagues as well. It's not uncommon for Zientarski to buttonhole an English teacher in the hallway and hand her a stack of the latest brain research — homework from the gym teacher.

It's because of their relentless spirit of investigation that I got to know these two men. Lawler heard me interviewed on the National Public Radio program *The Infinite Mind,* during which I referred to a protein that's elevated during exercise as "Miracle-Gro for the brain." Unbeknownst to me, Lawler began repeating the phrase in interviews of his own, including one with the director of a documentary film about obesity in America, *Super Size Me* (2004).

I had been looking for a concrete way to illustrate the effects of exercise on learning for this book, and focusing on a school district was a natural way to do that. But I also think the sheer size of the Naperville experiment gives it a broader resonance. The story is about students, but the lessons apply to adults too. What Naperville provides is a powerful case study on how aerobic activity can transform not only the body but also the mind. It also happens to be a wonderful template for reshaping our society.

So I made the journey to Illinois, and as I sat with Lawler and Zientarski in the atrium of the Naperville Holiday Inn, I listened to them say things I never expected to hear from a couple of coaches. "In our department, we create the brain cells," Zientarski says. "It's up to the other teachers to fill them."

A NEW STEREOTYPE: THE SMART JOCK

Lawler's tack runs opposite the trend in American public schools of cutting physical education in favor of increasing study time in math, science, and English — an effort to help students pass tests dictated by the No Child Left Behind Act. Only 6 percent of U.S. high schools offer a daily physical education class. At the same time, kids are spending an average of 5.5 hours a day in front of a screen of some sort — television, computer, or handheld device. It's not surprising that American children are less active than they've ever been.

That is why I was so inspired by what's going on in Naperville.

The first time I visited, it was just before school let out for the summer, but you wouldn't have known that by watching gym class at Madison Junior High. There must have been thirty kids jumping around with the sort of energy and enthusiasm you would only expect to see at the beginning of the school year: lining up to get on the climbing wall, arguing about who was going to get to use a new exercise bike attached to a video-game monitor, running wildly on treadmills, playing a video game called Dance Dance Revolution, where you dance on a control pad. They were all wearing heart rate monitors, and—most important—they were all *engaged*.

Some 30 percent of U.S. schoolchildren are overweight—six times more than in 1980—and another 30 percent are on the cusp. In Lawler's district, an astonishing 97 percent of freshmen in 2001, and again in 2002, were at a healthy weight according to body mass index guidelines from the Centers for Disease Control (CDC). In the spring of 2005, an independent assessment of Naperville 203 students' fitness showed even better results. A sports physiologist named Craig Broeder and a team of his graduate students from Benedictine University came in and tested a random sampling of 270 students, from sixth graders through high school seniors. "I can tell you that the Naperville school system is miles ahead of the national norm in terms of fitness," says Broeder, a former regional president of the American College of Sports Medicine. "It's not even close. They had *one* male out of a hundred thirty something who was obese. It's amazing. Their percentages of body fat were way below national norms using the CDC's height and weight standards. On other fitness variables, something like ninety-eight percent of the students passed."

Broeder is perfectly aware of Naperville's demographics, yet he's still impressed. "The numbers are too high for it to just be that," he says. "The PE program itself has to have had an additive impact on what that population would achieve otherwise. Let me put it this way: you can't say for sure that the PE program does it, but their fitness is so far off the scale that it can't be just because it's *Naperville*."

But what, exactly, do we know about the effect of gym on GPA? Few researchers have tackled the question, although one study from Virginia Tech showed that cutting gym class and allocating more time to math, science, and reading did not improve test scores, as so many school administrators assume it will. Because *gym class* can mean so many things, research in this area has focused on the correlation between physical fitness and academic achievement. The most telling studies come from the California Department of Education (CDE). Over the past five years, the CDE has consistently shown that students with higher fitness scores also have higher test scores.

The CDE correlated scores from standard achievement tests with scores from the FitnessGram, the state-mandated physical assessment, for more than one million students. The FitnessGram measures six areas: aerobic capacity, percentage of body fat, abdominal strength and endurance, trunk strength and flexibility, upper body strength, and overall flexibility. Students earn one point for each area if they pass the minimum requirements, so the top score on the FitnessGram is six. It's worth noting that this test doesn't measure *how* fit a student is, just *whether* he or she is acceptably fit in each area. In other words, it's pass-fail.

In 2001 fit kids scored twice as well on academic tests as their unfit peers. Among California's 279,000 ninth graders, for instance, those who scored a six on the FitnessGram ranked, on average, in the sixty-seventh percentile in math and the forty-fifth percentile in reading on the Stanford Achievement Test. If these scores seem less than stellar, consider those of the students who passed only one of the six areas: they ranked in the thirty-fifth and twenty-first percentiles, respectively.

When the CDE repeated the study in 2002, it factored in socioeconomic status. As expected, students with a higher standard of living scored better on the academic tests, but the results also showed that within the lower-income students, fitter kids scored better than unfit kids. This is a powerful statistic in itself. It suggests that although parents may not have immediate control over their financial situations, they can improve their kids' chances of performing well by encouraging them to get in shape. Exercise could break the cycle.

The California studies don't stand alone. In 2004 a panel of thirteen noted researchers in fields ranging from kinesiology to pediatrics conducted a massive review of more than 850 studies about the effects of physical activity on school-age children. Most of the studies measured the effects of thirty to forty-five minutes of moderate to vigorous physical activity three to five days a week. They covered a wide range of issues, such as obesity, cardiovascular fitness, blood pressure, depression, anxiety, self-concept, bone density, and academic performance. Based on strong evidence in a number of these categories, the panel issued a recommendation that schoolchildren should participate in one hour (or more) of moderate to vigorous physical activity a day. Looking specifically at academic performance, the panel found enough evidence to support the findings of the California studies, and it also reported that physical activity has a positive influence on memory, concentration, and classroom behavior. It didn't specify gym class, but you can see how the students in Naperville are getting a healthy jump start.

A WHOLE NEW BALLGAME

"I'm not a researcher; I'm a PE teacher," says Zientarski to a dozen educators packed into his cinderblock office at Naperville Central, as he hands them copies of the CDE studies. The educators come from a neighboring suburb, a school in South Side Chicago, as well as a rural district in Tulsa, Oklahoma, and they're here because Naperville 203 serves as a training academy for a nonprofit agency called PE4life, which has adopted the New PE philosophy. Illinois is the only state that requires daily phys ed, and PE4life is lobbying to change that—as well as the way it's taught. Zientarski stands up and announces, "Now, we're going to take a tour."

He leads the way, moving through the hallways with the deliberate stride of a seasoned U-boat commander. At the first stop, three student helpers are administering computerized health diagnostics to a group of sophomores with a computer system called TriFit. Giving the kids targets for heart rate, blood pressure, body fat, and the rest, he announces, is a proven method of motivating people to stay fit. Indeed, studies suggest that simply getting on the scale every morning improves the likelihood that someone who's overweight will shed pounds. But Lawler and Zientarski's ambitions extend far beyond concerns about their students' body mass index.

"I tell people it's not my job as a PE teacher to make kids fit," Zientarski says. "My job is to make them know all of the things they need to know to keep themselves fit. Exercise in itself is not fun. It's work. So if you can make them understand it, show them the benefits—that's a radical transformation. Especially for us coaches. We're control freaks. I can get sixty-five kids to stand on a white line if I say Hut!, and for years that's what we did."

Students in Naperville 203 had heart rate monitors before they had the Internet. When you walk into the gym at any of the district's schools today, it feels like you're in a state-of-the-art adult health club. Each has a TriFit assessment machine and weight machines, which in the junior highs are custom-made to accommodate students at that age. There are climbing walls and video-game-based aerobic machines. (Through Lawler's lobbying and Zientarski's browbeating, most of the equipment has been donated.)

The curriculum is designed to teach kids the principles, practice, and importance of fitness. When they reach high school, they're given a broad menu of options—from kayaking to dancing to rock climbing to typical team sports like volleyball and basketball—and shown how to draw up their own fitness plans. It's all centered around TriFit assessments students complete each year starting in fifth grade. They design their plans as freshmen and track their improvement until they graduate, at which time they get a fourteen-page health assessment. It combines fitness scores with factors like blood pressure and cholesterol levels, along with lifestyle and family history surveys, to predict risk of disease and suggest preventive measures. It is an astonishingly comprehensive document by any professional health standard, let alone one that an eighteen-year-old can carry in his hand as he steps into adult life. If only the rest of us could be so lucky.

Sports physiologist Craig Broeder, who conducted the fitness study in Naperville, remarks that students can choose from eighteen activities for gym. "One of the things that too many people forget is that you have to find something that allows a student to feel comfortable at excelling," he says. "So that it feels like *them* when they're doing it. When you only give a kid a limited option, like playing basketball, and you make it seem like punishment or boot camp, there's no way he's going to continue doing it. At Naperville, they give kids lots of options by which to excel; they design lifetime fitness activities." It's important for adults to remember this when considering how to get in shape.

Zientarski leads his group into the old girl's gym to show off the jewel of Central's physical education program: a twenty-four-foot-high, ninety-foot-long climbing wall and a high-ropes course they recently started using in a new leadership class. He gives an example of a drill he uses to teach trust and communication: the climber is blindfolded and has to rely on commands from his partner to reach the next hold on the wall. The newest part of the wall is set at an easier pitch for PE students with physical and mental disabilities. Answering the obvious concerns about liability, Zientarski says they have very few injuries in here because the kids are cooperating, not competing, and this is one of the most important lessons he and his staff teach.

"If you ask people, What is it you want our graduates to be able to know and do when they leave high school?" Zientarski explains, "They'll say, We want them to be able to communicate. We want them to be able to work in small groups. We want them to be able to problem solve. We want them to be risk takers. Where does that happen?" he asks, eyeballing his guests. "Science class? I don't think so."

GOOD FOR THE BODY, GOOD FOR THE BRAIN

About 135 miles south of Naperville, at the University of Illinois at Urbana-Champaign, a psychophysiologist named Charles Hillman conducted his own version of the CDE study with a group of 216 third and fifth graders and found the same correlation between fitness and academics. He and his coauthor, Darla Castelli, noticed something interesting. Of the six areas that the FitnessGram measures, two seem to be particularly important in relation to academic performance. "Body mass index and aerobic fitness really stuck out in our regression equation," Castelli says. "They were the most significant contributors. I was really surprised it was that clear-cut."

Hillman went beyond correlating data, though. He wanted to dig into the neuroscience of these findings, so he took a group of forty kids—half fit, half unfit—and measured their attention, working memory, and processing speed. During the cognitive testing, the kids wore something like a swim cap embedded with electrodes that measured electrical activity in the brain. The electroencephalogram (EEG) showed more activity in fit kids' brains, indicating that more neurons involved in attention were being recruited for a given task. "We see better integrity there," Hillman explains. In other words, better fitness equals better attention and, thus, better results.

Hillman also found something telling in how his subjects responded to making a mistake. While measuring their brain activity, he used what's called a flanker test, in which a series of five capital letters (Hs and Ss) are flashed on a screen. The only letter of interest is the one in the middle; the subject hits one button when it's an H and another button when it's an S. When something like HHSHH shows up, at the rate of once a second, it's easy to make a mistake, and you know as soon as you've done so. What Hillman found, he says, is that "fit kids slow down and make sure they get that next one right." The ability to stop and consider a response, to use the experience of a wrong choice as a guide in making the next decision, relates to executive function, which is controlled by an area of the brain called the prefrontal cortex. (I'll explore executive function in subsequent chapters, especially when we get to attention-deficit/hyperactivity disorder, which is partly caused by a lapse in the prefrontal cortex. If a child with ADHD took the flanker test, she would hit the wrong button before being able to stop herself, or hesitate too long to hit the right button. But you can imagine how much all of us rely on executive function.) Learning from our mistakes is profoundly important in everyday life, and Hillman's study shows that exercise — or at least the resulting fitness levels — can have a powerful impact on that fundamental skill.

FOLLOW THE LEADERS

There may be no better embodiment of Naperville's faith in the transformative power of exercise than Jessie Wolfrum. A self-described nerd and a straight-A student while at Central, she graduated in 2003 and enrolled at Embry-Riddle Aeronautical University in Daytona Beach, Florida, where she is now majoring in engineering physics. As a twin who tended to rely on her relationship with her sister rather than engage with other kids, Jessie had been shy all her life. "In third grade, my mom gave me the option of piano or soccer," Jessie recalls, laughing about it now. "I was so scared of the idea of hanging out with a bunch of girls at something I probably wasn't going to be good at that I picked something I didn't even like. I played piano for eight years!"

Of course, Phil Lawler didn't give her the piano option when she arrived as a student at Madison Junior High. Jessie had to participate, just like everyone else, and although she didn't much care for gym, it wasn't too terrible—certainly not traumatizing. And she learned lessons about her body that would serve her for years to come.

When she and her sister, Becky, moved on to Central, their divergent class schedules meant they weren't able to constantly lean on each other, so Jessie was forced to talk to other kids more often than she felt comfortable doing. She signed up for speech class to deal with her social anxiety, but she says what really helped her blossom was enrolling in kayaking. Jessie took to this skill-intensive sport immediately, and discovering she was good at something outside the academic realm helped transform her.

"If somebody notices that you're doing something that they can't do, you get some attention," Jessie says. "In kayaking, people started to notice me, and then I wasn't the person who faded into the wallpaper. It made me more adventurous. Even if you're shy, if somebody is like, How do you do that? you forget that you're shy, and you just explain it: you have to turn your head this way or do that with your paddle."

The swimming pool leveled the playing field in other ways too. "Once everybody changes into their swimsuits, you can't tell who's in the popular group," she says. "The class totally jumped those boundaries of social standings. I had a lot of problems with that until I took kayaking."

Emboldened by her experience in kayaking class, Jessie signed up for the leadership course being taught by Mr. Zientarski. The first thing he did was separate Jessie and her twin—and all of the other inseparable cliques. The leadership students learn to rock climb, and it's this sport in particular that captured Jessie's attention. She joined the Adventure Club, a sort of ad hoc Zero Hour for kids who wanted to come in at 6:30 in the morning to get extra time on the climbing wall or use the pool for kayaking.

Jessie and her sister actually decided to go paddling the morning of the Prairie State Achievement Examination, the Illinois version of the Scholastic Aptitude Test (SAT). They were so confident in their preparation, and so attuned to how exercise helped them focus, that they were comfortable splashing around a cold pool right before an important exam. How many high school kids do you know who would do that? How many *adults* do you know who would do that?

"When we showed up for the test, we were cold and wet," recalls Jessie. "We walked into the classroom, and we were the only ones who were awake. We ended up doing pretty well." They both scored 1400 out of a possible 1600—top-notch results.

When she got to college, Jessie continued pushing herself both academically and socially. She is an A/B student and, most surprisingly, she became a resident adviser, watching over a group of underclassmen on her hall, providing them with comfort, discipline, and counsel. She is no longer a wallflower.

It's tough to keep up with exercise in the transition from high school to college, but Jessie never strayed too far from her regimen. During her freshman year at Embry-Riddle, whenever anything stressful would come up, she and her roommate would run laps on the stairs in her dorm. That's something she learned back in Naperville—how to manage her brain with exercise. And that's the message I hope to deliver in this book.

"These days, every hour is sucked up with something—watching over residents, classes..." Jessie says. "When I don't have time to work out, I wish I did. Every time I know that a whole bunch of tests are coming up—when I'm really stressed out—I think, OK, you know how to handle this. It's definitely a relief to know that I have something to fall back on. If I didn't have that, I'd probably just go eat or something. But I know that exercise will spike up my brain activity, and so I think, Just go do it. I wouldn't know that if it weren't for my gym class."

BEYOND FITNESS

Like many people, I grew up thinking that gym was a joke. We had some fun, but, to my recollection, phys ed wasn't especially educational. As an adult, when I began lecturing to teachers and doctors about the positive impact of physical activity on mood, attention, self-esteem, and social skills, I certainly wasn't thinking of gym as the antidote. In my experience, PE wasn't really about exercise. Quite the opposite — it discouraged exercise. The cruel irony was that the shy, the clumsy, the out of shape — some of the kids who could most benefit from exercise — were pushed aside to sit on the bleachers. Someone like Jessie Wolfrum would have been marginalized and left to stew in her shame. Over the years, I've listened to a number of patients recount tales of humiliation in PE. The sidelines are fertile ground for developing the very sorts of issues that exercise ameliorates.

Part of the Naperville magic is that Lawler and Zientarski are exquisitely tuned in to this dynamic. "We used to do chin-ups," recalls Zientarski, with a tone bordering on disgust. "I would say about sixty-five percent of our boys couldn't do one chin-up. Come on down to PE class and be a failure!"

What strikes me about Zientarski's transformation from drill sergeant to sculptor of bodies, brains, and minds is how far he has been willing to go in redefining gym. For example, one of the most innovative changes he made at Central was to add a mandatory square-dancing class for freshman. It may not sound cutting edge, but the class is set up to use movement as a framework for teaching

29

social skills—a wonderful idea on many levels. In the first few weeks of the class, all the students receive scripts to use as conversation starters with their partners, and everyone switches partners after each dance. As the course progresses, the students are given time to interact without the scripts, first for thirty seconds and building up from there. The final exam is based on how accurately the students remember ten facts about a partner after spending fifteen minutes chatting.

Some kids who are socially timid never get a chance to learn how to talk to people and make friends, so they retreat, especially from the opposite sex. By not being singled out or relegated to a special social skills class, Zientarski's square-dance students get to practice how to talk and interact in a nontoxic setting. The activity serves both as a distraction and as a confidence builder. Some master the drill, and others merely break through their fears, but because everybody's doing it, it's less embarrassing.

When I talk to colleagues about the Naperville revolution and tell them that kids are learning these kinds of social skills in gym class, the reaction I get is stunned silence—they are in awe, just as I was. Throughout my work, I have spent a lot of time trying to identify and address the problems of what I call the social brain, and Zientarski has found the perfect prescription to help overcome the growing isolation and solitary nature of our lives today. In gym class! By having the structure, opportunity, and expectation, socially anxious students log in positive memories about the way to approach someone, how close to stand, and when to let the other person speak. Exercise serves as the social lubricant, and it's crucial to this kind of learning because it reduces anxiousness. Their brains are primed by the movement, and they lay down circuits that record the experience, which at first may be painful but which becomes less so in the context of an experience shared by the entire class. It's an intuitively brilliant way to bring kids out of their shells, at a poignant age when everyone feels self-conscious.

Zientarski puts them all in the same boat and gives them the tools and encouragement to build up their self-confidence. The dancing makes the whole lesson work.

It's offerings like this, I believe, that explain why so many parents in Naperville report that gym is their kids' favorite class. A mother named Olfat El-Mallak has two daughters who went to Madison and then Central. "It's not just physical exercise; it's something else that happens inside of them," she says. "This is almost like a motivational program. My girls believe in themselves. They are both very confident about themselves, and they didn't start this way. This is because of the PE program at District 203."

SPREADING THE GOSPEL

There are fifty-two million children, from kindergarten through twelfth grade, who attend public and private schools in the United States. If all of them had the benefit of Naperville-style physical education, our next generation of adults would be healthier, happier, and smarter. That is the ultimate goal of PE4life, the group that has hired Lawler to teach other educators the fitness-not-sports philosophy and methodology. About one thousand educators from 350 schools have been through the training, and many have since implemented their own versions of the program.

One such graduate is a man named Tim McCord. He is the physical education coordinator for the school district in Titusville, Pennsylvania, a defunct industrial town of six thousand that's been left for dead in a stretch of hill country between Pittsburgh and Lake Erie. This is where, back in 1859, the world's first successful oil well was drilled, but oil has come and gone, right along with the economy: The median income is now $25,000; 16 percent of the town is below the poverty line; and a few years back, about 75 percent of the kindergartners received government assistance for school lunches. Which is to say, this is not a wealthy suburb.

In 1999 McCord visited Naperville, came home, and transformed physical education in Titusville almost overnight. The district has twenty-six hundred students in one high school, one middle school, four elementary schools, and one early learning center. Titusville installed fitness centers in the secondary schools, bought heart rate monitors, and got the local hospital to help fund the TriFit diagnostics. Titusville even restructured the school day, adding ten minutes to the schedule and shaving time from *academic* classes to carve out time for daily gym. "It did not cost us a cent to do that," McCord says, noting that it was an administrator's suggestion. "And it's a huge move with No Child Left Behind—everybody else is going in the other direction."

Now Titusville's secondary schools have climbing walls, and the fitness centers are brimming with the latest training technology, most of it donated. The Cybex Trazer, for instance, is a brand-new device that looks like an upright computer station where students chase flashing lights. There are also cycling trainers, which allow kids to race one another on video screens or cue up routes from the Tour de France and compete with virtual Lance Armstrongs. McCord has also reached out to the community, opening the schools' fitness centers to members of the senior center. Within the schools, he's invited teachers in other subjects to get involved: English students use the heart rate monitors during public speaking, and math students use the data to learn how to graph.

Since the program started in 2000, the standardized test scores of Titusville's students have risen from below the state average to 17 percent above it in reading and 18 percent above it in math. Equally important are the psychosocial effects McCord has noticed: not a single fist fight among the 550 junior high kids since 2000. The district's bootstrap story has prompted visits from state representatives and even the president of the CDC. During one such show-and-tell, as McCord led a group past the junior high's climbing wall, he noticed a girl named Stephanie stuck about halfway up. Bookish and a little heavyset, she was on display for everyone to see her fail. But as her classmates noticed her struggling, they began cheering, "Go, Stephanie!" She made it to the top, and McCord spoke to her later. "She started to cry and couldn't believe the other kids were cheering her on," McCord recalls. "She said it helped her pull herself up."

The buzz about the broad effect of exercise on students is spreading among other government officials. Iowa Senator Tom Harkin recently held hearings about reestablishing physical education in schools based on news that one PE4life school in the inner city reduced its disciplinary problems by 67 percent. At Woodland Elementary School in Kansas City, Missouri, nearly all of the students have subsidized meal programs. In 2005 the physical education staff expanded gym from one class a week to forty-five minutes a day, focused almost entirely on cardiovascular activity. In the span of one school year, the students' fitness levels improved dramatically, and counselors reported that the number of incidents involving violence at Woodland decreased from 228 to 95 for the year.

For an inner-city school to go through such a rapid transformation, and for such a depressed town as Titusville to come alive as it has, is remarkable. McCord's community rallies around the Stephanies of the world rather than just the football team, and as the schoolchildren grow up, a larger percentage will continue to move and be active. They'll grab their kayak or bike instead of their Game Boy, and their minds and moods will be sharper for it.

Revolutions rely on youth, but as we've seen with Lawler, Zientarski, and McCord, even adults can make a major shift and recognize how physical activity influences the brain. If Titusville can find the spark, so can the rest of us. My hope is that we can use these examples as a new cultural model and, ultimately, reconnect the body and the brain. As you'll see, they belong together.

2
Learning
Grow Your Brain Cells

WHEN THE STUDENTS in Titusville or in Naperville go for a mile run in gym, they are more prepared to learn in their other classes: their senses are heightened; their focus and mood are improved; they're less fidgety and tense; and they feel more motivated and invigorated. The same goes for adults, in the classroom of life. What allows us to absorb the material is where the revolutionary new science comes into play. In addition to priming our state of mind, exercise influences learning directly, at the cellular level, improving the brain's potential to log in and process new information.

Darwin taught us that learning is the survival mechanism we use to adapt to constantly changing environments. Inside the microenvironment of the brain, that means forging new connections between cells to relay information. When we learn something, whether it's a French word or a salsa step, cells morph in order to encode that information; the memory physically becomes part of the brain. As a theory, this idea has been around for more than a century, but only recently has it been borne out in the lab. What we now know is that the brain is flexible, or *plastic* in the parlance of neuroscientists—more Play-Doh than porcelain. It is an adaptable organ that can be molded by input in much the same way as a muscle can be sculpted by lifting barbells. The more you use it, the stronger and more flexible it becomes.

The concept of plasticity is fundamental to understanding how the brain works and how exercise optimizes brain function by fostering that quality. Everything we do and think and feel is governed by how our brain cells, or neurons, connect to one another. What most people think of as psychological makeup is rooted in the biology of these connections. Likewise, our thoughts and behavior and environment reflect back on our neurons, influencing the pattern of connections. Far from being hardwired, as scientists once envisioned it, the brain is constantly being *re*wired. I'm here to teach you how to be your own electrician.

THE MEDIUM IS THE MESSENGER

It's all about communication. The brain is made up of one hundred billion neurons of various types that chat with one another by way of hundreds of different chemicals, to govern our every thought and action. Each brain cell might receive input from a hundred thousand others before firing off its own signal. The junction between cell branches is the synapse, and this is where the rubber meets the road. Synapses don't actually touch, which is a little confusing because neuroscientists talk about synapses "wiring together" when they establish a connection. The way it works is that an electrical signal shoots down the axon, the outgoing branch, until it reaches the synapse, where a neurotransmitter carries the message across the synaptic gap in chemical form. On the other side, at the dendrite, or the receiving branch, the neurotransmitter plugs into a receptor—like a key into a lock—and this opens ion channels in the cell membrane to turn the signal back into electricity. If the electrical charge at the receiving neuron builds up beyond a certain threshold, that nerve cell fires a signal along its own axon, and the entire process repeats.

About 80 percent of the signaling in the brain is carried out by two neurotransmitters that balance each other's effect: glutamate stirs up activity to begin the signaling cascade, and gamma-aminobutyric acid (GABA) clamps down on activity. When glutamate delivers a signal between two neurons that haven't spoken before, the activity primes the pump. The more often the connection is activated, the stronger the attraction becomes, which is what neuroscientists mean when they talk about binding. As the saying goes, neurons that fire together wire together. Which makes glutamate a crucial ingredient in learning.

Glutamate is a workhorse, but psychiatry focuses more on a group of neurotransmitters that act as regulators—of the signaling process and of everything else the brain does. These are serotonin, norepinephrine, and dopamine. And although the neurons that produce them account for only 1 percent of the brain's hundred billion cells, these neurotransmitters wield powerful influence. They might instruct a neuron to make more glutamate, or they might make the neuron more efficient or alter the sensitivity of its receptors. They can override other signals coming into the synapse, thus lowering the "noise" in the brain, or, conversely, amplify those signals. They can deliver signals directly, like glutamate and GABA, but their primary role is in adjusting the flow of information in order to fine-tune the overall balance of neurochemicals.

Serotonin, which you'll hear a lot more about in later chapters, is often called the policeman of the brain because it helps keep brain activity under control. It influences mood, impulsivity, anger, and aggressiveness. We use serotonin drugs such as fluoxetine (Prozac), for instance, because they help modify runaway brain activity that can lead to depression, anxiety, and obsessive-compulsiveness.

Norepinephrine, which was the first neurotransmitter scientists studied to understand mood, often amplifies signals that influence attention, perception, motivation, and arousal.

Dopamine, which is thought of as the learning, reward (satisfaction), attention, and movement neurotransmitter, takes on sometimes contradictory roles in different parts of the brain. Methylphenidate (Ritalin) eases attention-deficit/hyperactivity disorder (ADHD) by raising dopamine, thus calming the mind.

Most of the drugs we use to improve mental health target one or more of these three neurotransmitters. But as I hope to make abundantly clear, simply raising or lowering the level of a neurotransmitter doesn't elicit a crisp one-to-one result because the system is so complex. Manipulating just one neurotransmitter creates a ripple effect that takes different paths in different brains.

I tell people that going for a run is like taking a little bit of Prozac and a little bit of Ritalin because, like the drugs, exercise elevates these neurotransmitters. It's a handy metaphor to get the point across, but the deeper explanation is that exercise *balances* neurotransmitters—along with the rest of the neurochemicals in the brain. And as you'll see, keeping your brain in balance can change your life.

TO LEARN IS TO GROW

As fundamental as the neurotransmitters are, there's another class of master molecules that over the past fifteen years or so has dramatically changed our understanding of connections in the brain, specifically, how they develop and grow. I'm talking about a family of proteins loosely termed *factors*, the most prominent of which is brain-derived neurotrophic factor (BDNF). Whereas neurotransmitters carry out signaling, neurotrophins such as BDNF build and maintain the cell circuitry—the infrastructure itself.

During the 1990s, as neuroscientists began to pin down the cellular mechanism of memory, BDNF became the focus of a whole new field of research. About a dozen papers on BDNF were published before 1990, the year scientists discovered that it exists in the brain and nourishes neurons like fertilizer. Then, "a tsunami of labs and pharma companies" joined the fray, says Eero Castrén, a neuroscientist involved in the early work on BDNF at Sweden's Karolinska Institute. Today the research literature shows more than fifty-four hundred papers on BDNF. Once it became clear that BDNF was present in the hippocampus, an area of the brain related to memory and learning, researchers set out to test whether it's a necessary ingredient in the process.

Learning requires strengthening the affinity between neurons through a dynamic mechanism called long-term potentiation (LTP). When the brain is called on to take in information, the demand naturally causes activity between neurons. The more activity, the stronger the attraction becomes, and the easier it is for the signal to fire and make the connection. The initial activity marshals existing stores of glutamate in the axon to be sent across the synapse and reconfigures receptors on the receiving side to accept the signal. The voltage on the receiving side of the synapse becomes stronger in its resting state, thereby attracting the glutamate signal like a magnet. If the firing continues, genes inside the neuron's cell nucleus are turned on to produce more building material for the synapses, and it is this bolstering of the infrastructure that allows the new information to stick as a memory.

Say you're learning a French word. The first time you hear it, nerve cells recruited for a new circuit fire a glutamate signal between each other. If you never practice the word again, the attraction between the synapses involved naturally diminishes, weakening the signal. You forget. The discovery that astonished memory researchers—and earned Columbia University neuroscientist Eric Kandel a share of the 2000 Nobel Prize—is that repeated activation, or practice, causes the synapses themselves to swell and make stronger connections. A neuron is like a tree that instead of leaves has synapses along its dendritic branches; eventually new branches sprout, providing more synapses to further solidify the connections. These changes are a form of cellular adaptation called synaptic plasticity, which is where BDNF takes center stage.

Early on, researchers found that if they sprinkled BDNF onto neurons in a petri dish, the cells automatically sprouted new branches, producing the same structural growth required for learning—and causing me to think of BDNF as Miracle-Gro for the brain.

BDNF also binds to receptors at the synapse, unleashing the flow of ions to increase the voltage and immediately improve the signal strength. Inside the cell, BDNF activates genes that call for the production of more BDNF as well as serotonin and proteins that build up the synapses. BDNF directs traffic and engineers the roads as well. Overall, it improves the function of neurons, encourages their growth, and strengthens and protects them against the natural process of cell death. And—as I hope to make clear throughout this book—BDNF is a crucial biological link between thought, emotions, and movement.

THE MIND-BODY CONNECTION

Only a mobile creature needs a brain, points out New York University neurophysiologist Rodolfo Llinás in his 2002 book, *I of the Vortex: From Neurons to Self*. To illustrate, he uses the example of a tiny jellyfish-like animal called a sea squirt: Born with a simple spinal cord and a three hundred–neuron "brain," the larva motors around in the shallows until it finds a nice patch of coral on which to put down its roots. It has about twelve hours to do so, or it will die. Once safely attached, however, the sea squirt simply eats its brain. For most of its life, it looks much more like a plant than an animal, and since it's not moving, it has no use for its brain. Llinás's interpretation: "That which we call thinking is the evolutionary internalization of movement."

As our species has evolved, our physical skills have developed into abstract abilities to predict, sequence, estimate, plan, rehearse, observe ourselves, judge, correct mistakes, shift tactics, and then remember everything we did in order to survive. The brain circuits that our ancient ancestors used to start a fire are the same ones we use today to learn French.

Take the cerebellum, which coordinates motor movements and allows us to do everything from returning a tennis serve to resisting the pull of gravity. Starting with evidence that the trunk of nerve cells connecting the cerebellum to the prefrontal cortex are proportionally thicker in humans than in monkeys, it now appears that this motor center also coordinates thoughts, attention, emotions, and even social skills. I call it the rhythm and blues center. When we exercise, particularly if the exercise requires complex motor movement, we're also exercising the areas of the brain involved in the full suite of cognitive functions. We're causing the brain to fire signals along the same network of cells, which solidifies their connections.

When we learn something, a wide array of connected brain areas are called into action. The hippocampus doesn't do much without oversight from the prefrontal cortex. Broadly speaking, the prefrontal cortex organizes activity, both mental and physical, receiving input and issuing instructions through the brain's most extensive network of connections. The prefrontal cortex is the boss. As such, it is responsible for, among other things, keeping tabs on our current situation through so-called working memory, inhibiting stimuli and initiating action, judging, planning, predicting—all executive functions. As the CEO of the brain, the prefrontal cortex has to stay in close contact with the COO—the motor cortex—as well as many other areas.

The hippocampus is something like the cartographer, receiving new input from working memory, cross-referencing that information with existing memories for the sake of comparison and to form new associations, and reporting back to the boss. A memory, scientists believe, is a collection of information fragments dispersed throughout the brain. The hippocampus serves as a way station, receiving the fragments from the cortex, and then bundling them together and sending them back up as a map of a unique new pattern of connections.

Brain scans show that when we learn a new word, for example, the prefrontal cortex lights up with activity (as does the hippocampus and other pertinent areas, such as the auditory cortex). Once the circuit has been established by the firing of glutamate, and the word is learned, the prefrontal cortex goes dark. It has overseen the initial stages of the project, and now it can leave the responsibility to a team of capable employees while it moves on to new challenges.

This is how we come to know things and how activities like riding a bike become second nature. Patterns of thinking and movement that are automatic get stored in the basal ganglia, cerebellum, and brain stem—primitive areas that until recently scientists thought related only to movement. Delegating fundamental knowledge and skills to these subconscious areas frees up the rest of the brain to continue adapting, a crucial arrangement. Imagine if we had to stop and think to process every thought and to remember how to perform every action. We'd collapse in a heap of exhaustion before we could pour our first cup of morning coffee. Which is why a morning run is so important.

THE FIRST SPARK

In 1995 I was in the process of researching my book A User's Guide to the Brain, when I came across a one-page article in the journal Nature about exercise and BDNF in mice. There was scarcely more than a column of text, yet it said everything. Namely, that exercise elevates Miracle-Gro throughout the brain.

"I expected the big changes to occur in motor-sensory areas of the brain—the motor cortex, the cerebellum, the sensory cortex, maybe even the basal ganglia a little bit—because they're all involved with movement," recalls Carl Cotman, director of the Institute for Brain Aging and Dementia at the University of California, Irvine, who designed the study. "We developed the first films and, son of a gun, it showed up in the *hippocampus*. Well, the significance is that the hippocampus is an area of the brain that is extremely vulnerable to degenerative disease and that is needed for learning. Instantly I said, This changes the game completely."

The news certainly came out of left field for me. For years, I had been a vocal proponent of using exercise for ADHD and many other psychological issues, based on what I'd seen with my own patients and what I knew about exercise's effect on neurotransmitters. But this was different. By showing that exercise sparks the master molecule of the learning process, Cotman nailed down a direct biological connection between movement and cognitive function. In doing so, he blazed the trail for the study of exercise in neuroscience.

Cotman conducted this experiment not long after BDNF was discovered in the brain, and there was nothing to suggest that exercise had anything to do with it; his hypothesis was an act of sheer creativity. He'd just finished working on a long-term aging study designed to see if the people whose minds hold up best share anything in common. Among those with the least cognitive decline over a four-year period, three factors turned up: education, self-efficacy, and exercise. The first two weren't so surprising, but Cotman was curious about the last. "I got to thinking about what the heck was really going on," he says. "The assumption was that exercise didn't act on the brain, but my take on it was that somehow it had to be the brain."

At the time, if you'd asked what variable might underlie overall brain health, most scientists would have said neurotrophic factors because they were "kind of the in thing," says Cotman, and everyone knew that BDNF helped neurons in culture survive. It was a bit of a leap, but if Cotman could tie exercise to BDNF, he'd at least have a plausible explanation for why it turned up in the aging study.

He set up an experiment to measure the levels of BDNF in the brains of mice that exercise. It was important that the exercise be voluntary because if he forced the mice to run on treadmills, he feared his peers might say the effect was from the stress of being handled. No problem: he'd use running wheels. As an indication of how new this territory was, finding rodent equipment that the university would approve for lab use was an ordeal in itself—Cotman had to pay $1,000 apiece for stainless steel running wheels that would pass protocol. "I remember signing the purchase order and thinking, This is painful; I just hope it doesn't *not work*," he jokes. On top of that, none of his postdoctoral students wanted anything to do with this research, and he had to go through a number of graduate students before finding a physical therapy major who liked the idea.

Unlike humans, rodents seem to inherently enjoy physical activity, and Cotman's mice ran several kilometers a night. They were divided into four groups: mice running for two, four, or seven nights, and one control group with no running wheel. When their brains were injected with a molecule that binds to BDNF and scanned, not only did the scans of the running rodents show an increase in BDNF over controls, but the farther each mouse ran, the higher the levels were. When Cotman saw the results—that the spike occurred in the hippocampus—he didn't believe them himself: "I said, No, c'mon guys, we did something wrong; the darn hippocampus is lit up. We had to repeat the experiment—it was too far out. And so we did, and we got the same results."

As the stories of BDNF and exercise developed in parallel, it became clear that BDNF was important not merely for the survival of neurons but also for their growth (sprouting new branches) and thus for learning. Eero Castrén, as well as Susan Patterson from Kandel's lab at Columbia, found that if you stimulate LTP in mice by making them learn, BDNF levels increase. Looking inside their brains, researchers determined that mice without BDNF lose their capacity for LTP; conversely, injecting BDNF directly into the brains of rats encouraged LTP. Then one of Cotman's former post-doctoral students, neurosurgeon Fernando Gomez-Pinilla, showed that if you neutralize BDNF in mice, they are slow to find their way out of a pool having a hidden platform. It all adds up to solid evidence of how exercise helps the brain learn.

"One of the prominent features of exercise, which is sometimes not appreciated in studies, is an improvement in the *rate* of learning, and I think that's a really cool take-home message," Cotman says. "Because it suggests that if you're in good shape, you may be able to learn and function more efficiently."

Indeed, in a 2007 study of humans, German researchers found that people learn vocabulary words 20 percent faster following exercise than they did before exercise, and that the rate of learning correlated directly with levels of BDNF. Along with that, people with a gene variation that robs them of BDNF are more likely to have learning deficiencies. Without Miracle-Gro, the brain closes itself off to the world.

Psychiatry had grudgingly accepted the idea that exercise could help improve our state of mind by creating a conducive environment for learning. But Cotman's work laid the foundation for proving that exercise strengthens the cellular machinery of learning. BDNF gives the synapses the tools they need to take in information, process it, associate it, remember it, and put it in context. Which isn't to say that going for a run will turn you into a genius. "You can't just inject BDNF and be smarter," Cotman points out. "With learning, you have to respond to something in a different way. But the something has to be there."

And without question, what that something is matters.

THE NATURE OF NURTURE

Scientists all the way back to Ramón y Cajal—who won the Nobel Prize in 1906 for proposing that the central nervous system was made up of individual neurons that communicate at what he termed polarized junctions—have theorized that learning involves changes at the synapses. Despite the accolades, most scientists didn't buy it. And it wasn't until 1945 that a psychologist from McGill University named Donald Hebb stumbled onto the first hint of evidence. The lab rules were loose in those days, and apparently Hebb thought it would be fine if he brought some lab rats home as temporary pets for his children. The arrangement turned out to be mutually beneficial: When he returned the rats to the lab, Hebb noticed that compared to their cage-bound peers, they excelled in learning tests. The novel experience of being handled and toyed with somehow improved their learning ability, which Hebb interpreted to mean that it changed their brains. In his acclaimed textbook, *The Organization of Behavior: A Neuropsychological Theory*, he described the phenomenon as "use-dependent plasticity." The theory was that the synapses rearrange themselves under the stimulation of learning.

Hebb's work ties in with exercise because physical activity counts as novel experience, at least as far as the brain is concerned. In the 1960s a group of psychologists at Berkeley formalized an experimental model called environmental enrichment as a way to test use-dependent plasticity. Rather than take rodents home, the researchers outfitted their cages with toys, obstacles, hidden food, and running wheels. They also grouped the animals together, so they could socialize and play.

It wasn't all peace and love, though, and eventually the rodents' brains were dissected. Living in an environment with more sensory and social stimuli, the lab tests showed, altered the structure and function of the brain. Not only did the rats fare better on learning tasks, but their brains weighed more compared to those housed alone in bare cages. Hebb's definition of plasticity hadn't included growth. "This was at a time when it was almost heresy to say that the brain could actually change," says neuroscientist William Greenough—who, as a young graduate student during that period, was keenly interested in the Berkeley work—"especially in a physical way, through experience."

Greenough wanted to investigate environmental enrichment, but was warned off from that line of inquiry. "My adviser essentially said, If you pick that as your thesis, you'll be in Vietnam for sure," Greenough recalls. But as the Berkeley findings were replicated, the notion that experience could impact the brain gained a foothold. In a parallel line of research, a group from Harvard proved the converse—that environmental deprivation could *shrink* the brain. In examining cats raised with one eye sewn shut, they found that the visual cortex was significantly smaller. All this work established the metaphor of the brain as a muscle, and the notion of use it or lose it.

Aside from challenging the long-standing separation between biology and psychology, the social implications of environmental enrichment were radical. The Berkeley studies led to the creation of Head Start, the federal education program that provides funding to send disadvantaged children to preschool. Why should poor kids be left in bare cages? The field took off, and neuroscientists began to investigate different ways to stimulate brain growth.

Once Greenough was safely ensconced as a faculty member, at the University of Illinois, he turned back to this line of research. In a seminal study in the early 1970s, he used an electron microscope to show that environmental enrichment made the neurons sprout new dendrites. The branching caused by the environmental stimulation of learning, exercise, and social contact caused the synapses to form more connections, and those connections had thicker myelin sheaths, which allowed them to fire signals more efficiently.

Now we know that such growth requires BDNF. This remodeling of the synapses has a huge impact on the circuits' capacity to process information, which is profoundly good news. What it means is that you have the power to change your brain. All you have to do is lace up your running shoes.

STRETCHING PLASTICITY

As the concept of synaptic plasticity took hold in neuroscience, an even more radical notion of growth was gaining credence. For the better part of the twentieth century, scientific dogma held that the brain was hardwired once fully developed in adolescence, meaning we're born with all the neurons we're going to get. We can rearrange synapses all we like, but we can only lose neurons. Certainly, we can speed up the decline, a point that your eighth-grade biology teacher may have made to scare you away from underage drinking. "Now, remember: alcohol kills brain cells, and they *never* grow back."

But guess what? They *do* grow back—by the thousands. Not until scientists became handy with advanced imaging tools that enabled them to peer into the brain did they find conclusive evidence, which was published in a seminal 1998 paper. It came from an unlikely source. Cancer patients are sometimes injected with a dye that shows up in proliferating cells so that the spread of the disease can be tracked. Researchers looked at the brains of terminally ill patients who had donated their bodies to science and found that their hippocampi were packed with the dye marker, proof that neurons were dividing and propagating—a process called neurogenesis—just like cells in the rest of the body. With that, they formalized one of the biggest discoveries in neuroscience.

Ever since, from Stockholm to Southern California to Princeton, New Jersey, neuroscientists have been scrambling to figure out what our new brain cells actually do. The implications are wide-ranging, given that the fundamental cause of degenerative diseases such as Parkinson's and Alzheimer's is dying and damaged cells. Aging itself is a matter of cells dying, and suddenly we learned that the brain has a built-in countermeasure, at least in certain areas. Figure out how to kick-start neurogenesis, and maybe we could make replacement parts for the brain.

And what does this mean for healthy brains? One of the early clues about neurogenesis had come from studies of chickadees, which learn new songs every spring and also show a significant burst of new cells in the hippocampus. Coincidence? The fledgling cells hinted at some role in learning, but proof has been hard to come by. Like synaptic plasticity, "neurogenesis is clearly involved in our interactions with our environment, both emotionally and cognitively," says neuroscientist Fred Gage, of the Salk Institute in La Jolla, California. Gage was one of the researchers, along with Peter Eriksson of Sweden, who conducted the linchpin study in 1998. "Trying to figure out what exactly [neurogenesis] is doing is a real interesting problem."

Neurons are born as blank-slate stem cells, and they go through a development process in which they need to find something to do in order to survive. Most of them don't. It takes about twenty-eight days for a fledgling cell to plug into a network, and, as with existing neurons, Hebb's concept of activity-dependent learning would apply: if we don't use the newborn neurons, we lose them. Gage went back to the environmental enrichment model to test this idea in rodents. "When we first did our experiments, we had all sorts of things going on," Gage explains. "We needed to tease that out, and to our surprise, just putting a running wheel in a cage had a profound effect on the *number* of cells that were born. Ironically, with running, the same percentage of cells die as in the control group—it's just that you have a bigger starting pool. But in order for a cell to survive and integrate, it has to fire its axon." Exercise spawns neurons, and the stimulation of environmental enrichment helps those cells survive.

The first solid link between neurogenesis and learning came from one of Gage's colleagues, Henrietta van Praag. They used a rodent-size pool filled with opaque water to hide a platform just beneath the surface in one quadrant. Mice don't like water, so the experiment was designed to test how well they remember, from an earlier dip, the location of the platform—their escape route. When comparing inactive mice with others that hit the running wheel four to five kilometers a night, the results showed that the runners remembered where to find safety more quickly. Both groups swam at the same rate, but the exercised animals made a beeline for the platform, while the sedentary ones floundered about before figuring it out. When the mice were dissected, the active mice had twice as many new stem cells in the hippocampus as the inactive ones. Speaking generally about what they found, Gage says: "There is a significant correlation between the total number of cells and [a mouse's] ability to perform a complex task. And if you block neurogenesis, mice can't recall information."

Although all this research is in rodents, you can see how it might relate back to the kids in Naperville: Gym class provides the brain with the right tools to learn, and the stimulation in the kids' classes encourages those newly developing cells to plug into the network, where they become valuable members of the signaling community. The neurons are given a mission. And it seems that cells spawned during exercise are better equipped to spark LTP. They are plastic phenoms, which led Princeton neuroscientist Elizabeth Gould to suggest that perhaps our new neurons play a role in hanging onto our conscious thoughts, while the prefrontal cortex decides if they should be wired in as long-term memories. Gould is the researcher who first showed that primates grow new neurons, paving the way for experiments on human neurogenesis.

She and everyone else in the field of neuroscience are still unpacking the relationship between neurogenesis and learning, and exercise has been a crucial lab tool. What I find interesting, though, is that relatively few scientists are studying exercise because they're interested in exercise. Rather, they make the mice run because it "massively increases neurogenesis," as the title of a 2006 study in *Hippocampus* proclaimed, and thus allows researchers to deconstruct the chain of signals behind the process. That's what the pharmaceutical companies need to create drugs. They dream of an anti-Alzheimer's pill that regenerates neurons to keep memory intact. "There has to be some kind of chemical stuff in the [hippocampus] that is sensing exercise and saying, OK, let's start cranking out new cells," says Columbia University neurologist Scott Small, who recently used a novel MRI technique to track neurogenesis in live human subjects. "If we can identify those molecular pathways, we might be able to think of clever ways to induce neurogenesis biochemically."

Just imagine if they could put exercise in a bottle.

THE BODY-MIND CONNECTION

If we're going to have new cells, we'll need fertilizer for them, and from the get-go, neurogenesis researchers have been onto BDNF. They already knew that without Miracle Gro our brains can't take in new information, and now they've seen that BDNF is also a necessary ingredient for making new cells.

BDNF gathers in reserve pools near the synapses and is unleashed when we get our blood pumping. In the process, a number of hormones from the body are called into action to help, which brings us to a new list of initialisms: IGF-1 (insulin-like growth factor), VEGF (vascular endothelial growth factor), and FGF-2 (fibroblast growth factor). During exercise, these factors push through the blood-brain barrier, a web of capillaries with tightly packed cells that screen out bulky intruders such as bacteria. Scientists have just recently learned that once inside the brain, these factors work with BDNF to crank up the molecular machinery of learning.

They are also produced within the brain and promote stem-cell division, especially during exercise. The broader importance is that these factors trace a direct link from the body to the brain.

Take IGF-1, a hormone released by the muscles when they sense the need for more fuel during activity. Glucose is the major energy source for the muscles and the sole energy source for the brain, and IGF-1 works with insulin to deliver it to your cells. What's interesting is that the role of IGF-1 in the brain isn't related to fuel management, but to learning—presumably so we can remember where to locate food in the environment. During exercise, BDNF helps the brain increase the uptake of IGF-1, and it activates neurons to produce the signaling neurotransmitters, serotonin and glutamate. It then spurs the production of more BDNF receptors, beefing up connections to solidify memories. In particular, BDNF seems to be important for long-term memories.

Which makes perfect sense in light of evolution. If we strip everything else away, the reason we need an ability to learn is to help us find and obtain and store food. We need fuel to learn, and we need learning to find a source of fuel—and all these messengers from the body keep this process going and keep us adapting and surviving.

To pipe fuel to new cells, we need new blood vessels. When our body's cells run short of oxygen, as they can when our muscles contract during exercise, VEGF gets to work building more capillaries in the body and the brain. Researchers suspect that one way VEGF is vital to neurogenesis is its role in changing the permeability of the blood-brain barrier, prying back the fence to let other factors through during exercise.

Another important element from the body that makes its way to the brain is FGF-2, which, like IGF-1 and VEGF, is increased during exercise and is necessary for neurogenesis. In the body, FGF-2 helps tissue grow, and in the brain it's important to the process of LTP.

As we age, production of all three of these factors and BDNF naturally tails off, bringing down neurogenesis with it. Even before we get old, however, a drop in these factors and in neurogenesis can show up in stress and depression, as we'll see later. To me, this is actually encouraging news, because if moving the body increases BDNF, IGF-1, VEGF, and FGF-2, it means we have some control over the situation.

It's about growth versus decay, activity versus inactivity. The body was designed to be pushed, and in pushing our bodies we push our brains too. Learning and memory evolved in concert with the motor functions that allowed our ancestors to track down food, so as far as our brains are concerned, if we're not moving, there's no real need to learn anything.

EXERCISE YOUR OPTIONS

Now you know how exercise improves learning on three levels: first, it optimizes your mind-set to improve alertness, attention, and motivation; second, it prepares and encourages nerve cells to bind to one another, which is the cellular basis for logging in new information; and third, it spurs the development of new nerve cells from stem cells in the hippocampus. OK, but now you want to know what the best exercise plan is. I wish there were an ideal type and amount of activity to suggest for building your brain, but scientists are only beginning to tackle such questions. "Nobody's done that research yet," says William Greenough. "But I suspect in five years we'll know a lot more."

Still, we can draw certain conclusions from the existing research. One thing scientists know for sure is that you can't learn difficult material *while* you're exercising at high intensity because blood is shunted away from the prefrontal cortex, and this hampers your executive function. For example, while working out on the treadmill or the stationary bike for twenty minutes at a high intensity of 70 to 80 percent of their maximum heart rate, college students perform poorly on tests of complex learning. (So don't study for the Law School Admission Test with the elliptical machine on full-tilt.) However, blood flow shifts back almost immediately after you finish exercising, and this is the perfect time to focus on a project that demands sharp thinking and complex analysis.

A notable experiment in 2007 showed that cognitive flexibility improves after just one thirty-five-minute treadmill session at either 60 percent or 70 percent of maximum heart rate. The forty adults in the study (age fifty to sixty-four) were asked to rattle off alternative uses for common objects, like a newspaper — it's meant for reading, but it can be used to wrap fish, line a birdcage, pack dishes, and so forth. Half of them watched a movie and the other half exercised, and they were tested before the session, immediately after, and again twenty minutes later. The movie watchers showed no change, but the runners improved their processing speed and cognitive flexibility after just one workout. Cognitive flexibility is an important executive function that reflects our ability to shift thinking and to produce a steady flow of creative thoughts and answers as opposed to a regurgitation of the usual responses. The trait correlates with high-performance levels in intellectually demanding jobs. So if you have an important afternoon brainstorming session scheduled, going for a short, intense run during lunchtime is a smart idea.

A lot of the research I've mentioned in this chapter revolves around exercise's effect on the hippocampus, because its role in forming memories makes it vital to learning. But the hippocampus isn't off by itself somewhere, stamping out new circuits on its own accord. The learning process calls on a lot of areas, under the direction of the prefrontal cortex. The brain has to be aware of the incoming stimulus, hold it in working memory, give it emotional weight, associate it with past experience, and relate all this back to the hippocampus. The prefrontal cortex analyzes the information, sequences it, and ties everything together. It works with the cerebellum and the basal ganglia, which keep these functions on track by maintaining rhythm for the back-and-forth of information. Improving plasticity in the hippocampus strengthens a crucial link in the chain, but learning creates bushier, healthier, better connected neurons throughout the brain. The more we build these networks and enrich our stores of memory and experience, the easier it is to learn, because what we already know serves as a foundation for forming increasingly complex thoughts.

As for how much aerobic exercise you need to stay sharp, one small but scientifically sound study from Japan found that jogging thirty minutes just two or three times a week for twelve weeks improved executive function. But it's important to mix in some form of activity that demands coordination beyond putting one foot in front of the other. Greenough worked on an experiment several years ago in which running rats were compared to others that were taught complex motor skills, such as walking across balance beams, unstable objects, and elastic rope ladders. After two weeks of training, the acrobatic rats had a 35 percent increase of BDNF in the cerebellum, whereas the running rats had none in that area. This extends what we know from the neurogenesis research: that aerobic exercise and complex activity have different beneficial effects on the brain. The good news is they're complementary. "It's important to take both into account," says Greenough. "The evidence isn't perfect, but really, your regimen has to include skill acquisition and aerobic exercise."

What I would suggest, then, is to either choose a sport that simultaneously taxes the cardiovascular system and the brain—tennis is a good example—or do a ten-minute aerobic warm-up before something nonaerobic and skill-based, such as rock climbing or balance drills. While aerobic exercise elevates neurotransmitters, creates new blood vessels that pipe in growth factors, and spawns new cells, complex activities put all that material to use by strengthening and expanding networks. The more complex the movements, the more complex the synaptic connections. And even though these circuits are created through movement, they can be recruited by other areas and used for thinking. This is why learning how to play the piano makes it easier for kids to learn math. The prefrontal cortex will co-opt the mental power of the physical skills and apply it to other situations.

Learning the asanas of yoga, the positions of ballet, the skills of gymnastics, the elements of figure skating, the contortions of Pilates, the forms of karate — all these practices engage nerve cells throughout the brain. Studies of dancers, for example, show that moving to an irregular rhythm versus a regular one improves brain plasticity. Because the skills involved in these activities are unnatural forms of movement, they serve as activity-dependent learning of the sort that made Hebb's rats smarter and that Greenough showed made synapses bushier.

Any motor skill more complicated than walking has to be learned, and thus it challenges the brain. At first you're awkward and flail a little bit, but then as the circuits linking the cerebellum, basal ganglia, and prefrontal cortex get humming, your movements become more precise. With the repetition, you're also creating thicker myelin around the nerve fibers, which improves the quality and the speed of the signals and, in turn, the circuit's efficiency. To take the example of karate, as you perfect certain forms, you can incorporate them into more complicated movements, and before long you have new responses to new situations. The same would hold true for learning tango. The fact that you have to react to another person puts further demands on your attention, judgment, and precision of movement, exponentially increasing the complexity of the situation. Add in the fun and social aspect, and you're activating the brain and the muscles all the way down through the system. And then you're primed and ready to move on to the next challenge, which is what it's all about.

3

Stress

The Greatest Challenge

SUSAN WAS STRESSED out. It had been more than a year since the remodeling contractor had taken over her kitchen, yet she had begun to fear the interludes of silence more than the construction racket itself. Silence meant work had stopped, for whatever reason, and that meant the job would take even longer. She had no idea when she would get her kitchen back, let alone her life. It was terribly unsettling, as anyone who has survived a remodeling project can attest: strangers traipsing in and out all day, no control over your time, sheetrock dust everywhere—utter confusion. All the while, the contractor himself seems perfectly at home in your house, if and when he shows up.

A woman in her forties, Susan had always been an active, out-going person: mother of three school-age boys, head of the PTA, an equestrian, a professional volunteer with a full schedule of com-mitments. But suddenly she was forced to stick around the house all day, waiting for the workmen to arrive, often only to have them cancel. It was enough to drive anyone batty. She was a shut-in in her own ripped-up home, and she didn't know what to do with herself. To take the edge off, she started having a glass of wine. And then another. Before long, she found herself uncorking a nice chardonnay before lunchtime. "Always chardonnay," she says. "It's the only thing I drink."

Susan's world was shrinking, and so too, as I'll explain, was her brain. She came in to see me because she was worried that her coping mechanism was becoming an addiction. As she sat in my office, we discussed ways she could break the cycle of reaching for the wine whenever she felt stressed. I wanted to help her find something she could do immediately, right at home, first to distract her, but also to relieve the feeling of stress. She wasn't much for the gym, but she was fairly athletic, and somehow it came out that she liked to jump rope. Perfect. I suggested she start jumping rope every time she felt the stress coming on.

The next time I saw her, she told me she had jump ropes stashed on different floors of the house, and she'd been able to stop using wine to relieve her stress. Even with just those short bursts of activity, she immediately felt more in control, like the master of her own fate. She also felt a genuine relief — less tension in her muscles and less distracting activity in her mind. She explains it this way: "I feel like it kind of reboots my brain."

REDEFINING STRESS

Everyone knows stress, but do they really? It comes in many shapes and sizes, acute and chronic — social stress, physical stress, metabolic stress, to name a few. Most people use the word indiscriminately for both cause and effect. That is, the stress the world exerts upon us — "There is a lot of stress at work right now" — as well as the feeling we get inside when everything seems like too much: "I'm so stressed, I can't think straight." Scientists themselves don't always distinguish between the psychological state of stress and the physiological response to stress.

Stress is such a malleable term partly because the feeling can span a wide emotional range, from a mild state of alertness to a sense of being completely overwhelmed by the push and pull of life. At the far end of the spectrum is what you know as being stressed out—a lonely place where issues that might ordinarily seem like challenges take on the proportions of insurmountable problems. Stay there too long, and we're talking about chronic stress, which translates emotional strain into physical strain. This is where the ripple effects of the body's stress response can lead to full-blown mental disorders such as anxiety and depression, as well as high blood pressure, heart problems, and cancer. Chronic stress can even tear at the architecture of the brain.

But how to make sense of such a woolly concept as stress? By keeping in mind its biological definition. Above all, stress is a threat to the body's equilibrium. It's a challenge to react, a call to adapt. In the brain, anything that causes cellular activity is a form of stress. For a neuron to fire, it requires energy, and the process of burning fuel creates wear and tear on the cell. The feeling of stress is essentially an emotional echo of the underlying stress on your brain cells.

You probably wouldn't think of getting out of a chair as stressful—it doesn't *feel* stressful—but, biologically speaking, it most definitely is. It doesn't compare with, say, losing a job, but here's the thing: Both events activate parts of the same pathways in the body and the brain. Standing up triggers neurons needed to coordinate the movement, and dreading unemployment generates plenty of activity, since emotions are a product of neurons signaling one another. Likewise, learning French, meeting new people, and moving your muscles all make demands on your brain; all are forms of stress. As far as your brain is concerned, stress is stress—the difference is in degree.

INOCULATE YOURSELF

How the body and brain respond to stress depends on many factors, not the least of which is your own genetic background and personal experience. Today there is an ever-widening gap between the evolution of our biology and our society. We don't have to run from lions, but we're stuck with the instinct, and the fight-or-flight response doesn't exactly fly in the boardroom. If you get stressed at work, would you slap your boss? Or turn and run? The trick is how you respond. The way you choose to cope with stress can change not only how you feel, but also how it transforms the brain. If you react passively or if there is simply no way out, stress can become damaging. Like most psychiatric issues, chronic stress results from the brain getting locked into the same pattern, typically one marked by pessimism, fear, and retreat. Active coping moves you out of this territory. Instincts aside, you *do* have some control over how stress affects you. And, as Susan would agree, control is key.

Exercise controls the emotional and physical feelings of stress, and it also works at the cellular level. But how can that be, if exercise itself is a form of stress? The brain activity caused by exercise generates molecular by-products that can damage cells, but under normal circumstances, repair mechanisms leave cells hardier for future challenges. Neurons get broken down and built up just like muscles — stressing them makes them more resilient. This is how exercise forces the body and mind to adapt.

Stress and recovery. It's a fundamental paradigm of biology that has powerful and sometimes surprising results.

In the 1980s, the U.S. Department of Energy (DOE) commissioned a study on the health impacts of sustained radiation exposure. They compared two groups of nuclear shipyard workers from Baltimore who had similar jobs except for a single key difference: one group was exposed to very low levels of radiation from the materials they handled, and the other was not. The DOE tracked the workers between 1980 and 1988, and what they found shocked everyone involved.

Radiation made them healthier. The twenty-eight thousand workers exposed to radiation had a 24 percent lower mortality rate than their thirty-two thousand counterparts who were not exposed to radiation. Somehow, the toxins that everyone assumed and feared were ruining the workers' health were doing just the opposite. Radiation is a stress in that it damages cells, and at high levels it kills them and can lead to the development of diseases such as cancer. In this case, the radiation dose was apparently low enough that instead of killing the cells of the exposed workers, it made them stronger.

Maybe stress isn't so bad after all. But because the study "failed"— it didn't show the expected malignant effect of radiation—it was never published. From what we've since learned about the biology of stress and recovery, stress seems to have an effect on the brain similar to that of vaccines on the immune system. In limited doses, it causes brain cells to overcompensate and thus gird themselves against future demands. Neuroscientists call this phenomenon stress inoculation.

What's gotten lost amid all the advice about how to reduce the stress of modern life is that challenges are what allow us to strive and grow and learn. The parallel on the cellular level is that stress sparks brain growth. Assuming that the stress is not too severe and that the neurons are given time to recover, the connections become stronger and our mental machinery works better. Stress is not a matter of good and bad—it's a matter of *necessity*.

THE ALARM SYSTEM

Triggered by a primitive call to survive, the body's stress response is a built-in gift of evolution without which we wouldn't be here today. The response ranges from mild to intense depending on the cause. Severe stress activates the emergency phase, commonly known as the fight-or-flight response. It's a complex physiological reaction that marshals resources to mobilize body and brain, and engraves a memory of what happened, so we can avoid it next time. *Where was that lion, exactly?* The threat has to be fairly intense for the body to get involved, but any degree of stress activates fundamental brain systems—those that manage attention, energy, and memory. If we strip away everything else, our ingrained reaction to stress is about focusing on the danger, fueling the reaction, and logging in the experience for future reference, which I think of as wisdom. It is only in recent years that scientists have begun to recognize and describe the role of stress in the formation and recall of memories. The development of this understanding is exciting because it sheds light on why—(and how)—stress can have such a profound effect on the way we perceive the world.

The fight-or-flight response calls into action several of the body's most powerful hormones and scores of neurochemicals in the brain. The brain's panic button, called the amygdala, sets off the chain reaction on receiving sensory input about a possible threat to the body's natural equilibrium. Being hunted would certainly qualify, but so would being the hunter. The amygdala's job is to assign intensity to the incoming information, which may or may not be obviously survival related. It's not just about fear, but any intense emotional state, including, for example, euphoria or sexual arousal. Winning the lottery or dining with a supermodel can trigger the amygdala. These events may not seem stressful, but remember, our brains don't distinguish between "good" and "bad" demands on the system. And in an evolutionary light, good fortune and a good date are related to survival—prospering and procreating.

The amygdala connects to many parts of the brain and thus receives a wide array of input—some of it routed through the high-level processing center of the prefrontal cortex, and some of it wired indirectly, bypassing the cortex, which explains how even a subconscious perception or memory can trigger a stress response.

Within ten milliseconds of sounding the alarm, the amygdala fires off messages that cause the adrenal gland to release different hormones at different stages. First, norepinephrine triggers lightning-fast electrical impulses that travel through the sympathetic nervous system activate the adrenal gland to dump the hormone epinephrine, or adrenaline, into the bloodstream. Heart rate, blood pressure, and breathing increase, contributing to the physical agitation we feel under stress. At the same time, signals carried by norepinephrine and corticotropin-releasing factor (CRF) travel from the amygdala to the hypothalamus, where they are handed off to messengers that take the slow train through the bloodstream. These messengers prompt the pituitary gland to activate another part of the adrenal gland, which releases the second major hormone of the stress response: cortisol. This relay from the hypothalamus to the pituitary to the adrenal gland is known as the HPA axis, and its role in summoning cortisol and in turning off the response makes it a key player in the story of stress. Meanwhile, the amygdala has signaled the hippocampus to start recording memories and another dispatch has been sent to the prefrontal cortex, which decides whether the threat truly merits a response.

Humans are unique among animals in that the danger doesn't have to be clear and present to elicit a response — we can anticipate it; we can remember it; we can conceptualize it. And this capacity complicates our lives dramatically. "The mind is so powerful that we can set off the [stress] response just by imagining ourselves in a threatening situation," writes Rockefeller University neuroscientist Bruce McEwen in his book *The End of Stress as We Know It*. In other words, we can think ourselves into a frenzy.

There is an important flip side to McEwen's point: We can literally run ourselves *out* of that frenzy. Just as the mind can affect the body, the body can affect the mind. But the idea that we can alter our mental state by physically moving still has yet to be accepted by most physicians, let alone the broader public. It's a fundamental theme of my work, and it's particularly relevant in the context of stress. After all, the purpose of the fight-or-flight response is to mobilize us to act, so physical activity is the natural way to prevent the negative consequences of stress. When we exercise in response to stress, we're doing what human beings have evolved to do over the past several million years. On one level, it's that simple. Of course, there are many levels to explore.

FOCUS

The overarching principle of the fight-or-flight response is marshaling resources for immediate needs in lieu of building for the future—act now, ask questions later. The hormonal rush of epinephrine focuses the body, increasing heart rate and blood pressure and dilating the bronchial tubes of the lungs to carry more oxygen to the muscles. Epinephrine binds to muscle spindles, and this ratchets up the muscles' resting tension so they're ready to explode into action. Blood vessels in the skin constrict to limit bleeding in the event of a wound. Endorphins are released in the body to blunt pain. In this scenario, biological imperatives like eating and reproduction are put on the back burner. The digestive system shuts down; the muscles used to contract the bladder relax so as not to waste glucose; and saliva stops flowing.

If you've ever faced a nerve-racking public-speaking situation, you've experienced this shift in the form of a racing heart and cotton mouth. Your muscles and your brain get stiff, and you lose all hope of being flexible and engaging. Or, if the processed signal from the cortex to the amygdala breaks up, you can't think and you freeze. Technically, the full-blown stress response should be called "freeze or fight or flight." None of this is particularly helpful when you're up at the podium, but the body responds in essentially the same way whether you're staring down a hungry lion or a restless audience.

Two neurotransmitters put the brain on alert: norepinephrine arouses attention, then dopamine sharpens and focuses it. An imbalance of these neurotransmitters is why some people with attention-deficit/hyperactivity disorder (ADHD) come across as stress junkies. They have to get stressed to focus. It's one of the primary factors in procrastination. People learn to wait until the Sword of Damocles is ready to fall—it's only then, when stress unleashes norepinephrine and dopamine, that they can sit down and do the work. A need for stress also explains why ADHD patients sometimes seem to shoot themselves in the foot. When everything is going well, they need to stir up the situation, and they subconsciously find a way to create a crisis. I have one patient who, after a series of dysfunctional relationships, finally found a guy she really admires and who treats her well. Yet every time things are good, she picks a fight. Reminding her of the stress-junkie pattern helps her to be aware of her tendencies and—I hope—catch herself before she starts trouble.

FUEL

To fuel the anticipated activity of the muscles and the brain, epinephrine immediately begins converting glycogen and fatty acids into glucose. Traveling through the bloodstream, cortisol works more slowly than epinephrine, but its effects are incredibly widespread. Cortisol wears a number of different hats during the stress response, one of which is that of traffic cop for metabolism. Cortisol takes over for epinephrine and signals the liver to make more glucose available in the bloodstream, while at the same time blocking insulin receptors at nonessential tissues and organs and shutting down certain intersections so the fuel flows only to areas important to fight-or-flight. The strategy is to make the body insulin-resistant so the brain has enough glucose. Cortisol also begins restocking the shelves, so to speak, replenishing energy stores depleted by the action of epinephrine. It converts protein into glycogen and begins the process of storing fat.

If this process continues unabated, as in chronic stress, the action of cortisol amasses a surplus fuel supply around the abdomen in the form of belly fat. (Unrelenting cortisol also explains why some marathon runners carry a slight paunch despite all their training—their bodies never get a chance to adequately recover.) The problem with our inherited stress response is that it mobilizes energy stores that don't get used. More on that later.

During the initial phase of the stress response, cortisol also spurs the release of insulin-like growth factor (IGF-1), which is a crucial link in fueling the cells. The brain is a conspicuous consumer of glucose, using 20 percent of the available fuel even though it accounts for only about 3 percent of our body weight. But it has no capacity to store fuel, so cortisol's role in providing a steady flow of glucose is critical to proper brain function. Operating on a fixed budget of fuel, the brain has evolved to shift energy resources as necessary, meaning that mental processing is competitive. It's simply not possible to have all of our neurons firing at once, so if one structure is active, it must come at the expense of another. One of the problems with chronic stress is that if the HPA axis is guzzling all the fuel to keep the system on alert, the thinking parts of the brain are being robbed of energy.

WISDOM

Recording memories of stressful situations is an adaptive behavior with obvious evolutionary benefits. It's the wisdom of our collective experience that has allowed us to survive, and cortisol has played a major role. Neuroendocrinologist Bruce McEwen first found cortisol receptors in the hippocampus in rat brains in the 1960s, and then in rhesus monkeys, and now we know they exist in humans. The finding initially alarmed scientists because the stress hormone had been shown to be toxic to brain cells in a petri dish. "What is cortisol actually doing to affix these memories?" he asks. "All we can say is that if you don't have adequate cortisol receptors in the hippocampus at the time these memories are being formed, the learning process is less efficient. The exact details are still being worked out."

It seems that, like stress itself, cortisol isn't simply good or bad. A little bit helps wire in memories; too much suppresses them; and an overload can actually erode the connections between neurons and destroy memories. The hippocampus provides context for the memory—the what, how, where, and when—and the amygdala provides the emotional content—the fear or the excitement. With direction from the prefrontal cortex, the hippocampus can compare memories and say, "Don't worry; it's a stick, not a snake," and so has the capacity to shut off the HPA axis directly and put an end to the stress response. As long as it's not overexcited.

Within minutes of the alarm bell, cortisol, CRF, and norepinephrine—the major stress agents in the brain—bind to cell receptors that boost glutamate, the excitatory neurotransmitter responsible for all of the signaling in the hippocampus. Increasing glutamate activity speeds up the flow of information in the hippocampus and changes the dynamic at the synapses, so that each time a message is sent, the signal fires more easily, thus requiring less glutamate. Initially, then, the stress response enhances long-term potentiation (LTP), the fundamental mechanism of memory.

Short-term memory probably results from this initial increase in the excitability of hippocampal neurons. Then, as levels of cortisol peak, cortisol turns on genes inside the cells that make more proteins used as building material for cells: more dendrites, more receptors, and bulkier synapses. This is where things get curious. The beefed-up cells cement the survival memory and shield the neurons in that circuit from other demands. A neuron might be part of any number of memories. But if a potential memory comes along during stress, it has a more difficult time recruiting neurons to be part of its own new circuit. It needs to clear a certain threshold to make an impression.

This likely explains why memories not related to the stressor are blocked during the stress response. It also helps explain why constantly high levels of cortisol—due to chronic stress—make it hard to learn new material, and why people who are depressed have trouble learning. It's not just lack of motivation, it's because the hippocampal neurons have bolstered their glutamate machinery and shut out less important stimuli. They're obsessed with the stress.

Human studies also show that excess cortisol can block access to *existing* memories, which explains how people can forget where the fire exit is when there's actually a fire—the lines are down, so to speak. With too much stress, we lose the ability to form unrelated memories, and we might not be able to retrieve the ones we have. The next time you're forced to participate in a fire drill, consider that the neurological point of the exercise is to make those circuits stronger, to burn in the memory. With an overload of stress, as I'll explain later, you get the petri dish effect—cortisol eroding neurons.

FIGHTING OUR INSTINCTS

The stress response is elegantly adaptive behavior, but because it doesn't get you very far in today's world, there's no outlet for all that energy buildup. You have to make a conscious effort to initiate the physical component of fight or flight.

The human body is built for regular physical activity, but how much? In a 2002 article in the *Journal of Applied Physiology*, researchers studied this very question, by looking at our ancestors' pattern of physical activity, which they call the Paleolithic rhythm. From the time *Homo sapiens* emerged two million years ago, until the agricultural revolution, ten thousand years ago, everyone was a hunter-gatherer, and life was marked by periods of intense physical activity followed by days of rest. It was feast or famine. By calculating how much our forebears "exercised" and comparing it to figures from today, it's easy to see where the problem lies: Our average energy expenditure per unit of body mass is less than 38 percent of that of our Stone Age ancestors. And I think it's fair to say that our calorie intake has increased quite a bit. The kicker is that even if we followed the most demanding governmental recommendations for exercise and logged thirty minutes of physical activity a day, we'd still be at less than half the energy expenditure for which our genes are encoded. Paleolithic man had to walk five to ten miles on an average day, just to be able to eat.

Today we don't have to expend much energy to find food, and we certainly don't have to use our brains to figure out how to get our next meal. This situation has come about only in the past century or so, but it takes tens of thousands of years for our biology to evolve—there's a mismatch between our lifestyle and our genes. Human genes are thrifty by nature, so we end up stockpiling calories while we're sitting at our desks.

In the context of stress, the great paradox of the modern age may be that there is not more hardship, just more news—and too much of it. The 24/7 streaming torrent of tragedy and demands flashing at us from an array of digital displays keeps the amygdala flying. The negative and the hectic and the hopeless heap on the stress, but we figure we can handle it because we always have. Up to a point. Then, we just want to relax and take a break, so we grab a drink and plop down in front of the TV or go sit on a beach somewhere. It's no wonder that obesity has doubled in the past twenty years—our lifestyle today is both more stressful and more sedentary.

Maybe you've seen ads for drugs that slim down the belly by blocking cortisol. The belly is just doing its job, stockpiling energy stores as insurance against the next famine. With chronic stress, that stockpile ends up around the midsection, in the form of a spare tire. This is detrimental not only to our physique, but also to our health, because fat stores can easily make their way into the arteries of the heart and cause blockage. For anyone skeptical of the notion that stress can kill, herein lies one of the physical links between stress and heart attacks.

Compounding the buildup of fat, after a stressful event, we often crave comfort foods. Our body is calling for more glucose, and simple carbohydrates and fat—like those glistening in a box of Dunkin' Donuts—are readily converted into fuel. And in modern life, people tend to have fewer friends and less support, because there's no tribe. Being alone is not good for the brain.

A common protocol scientists use to induce the physiological stress response in rats is to remove them from their social structure; simply isolating them activates stress hormones. The same is true in humans: It's stressful to be shunned or isolated. Loneliness is a threat to survival. Not coincidentally, the less physically active we are, the less likely we'll be to reach out and touch someone. Studies show that by adding physical activity to our lives, we become more socially active—it boosts our confidence and provides an opportunity to meet people. The vigor and motivation that exercise brings helps us establish and maintain social connections.

There's nothing wrong with your desire to take a break. The issue is how you choose to spend that time. It's the comfort foods, the quick fats and sugars, the alcohol to take the edge off, or, for some people, drugs or other addictions that cause problems. If you exercise or even just socialize, you're tapping into the evolutionary antidote to stress.

Sometimes it's a simple matter of substitution, as my patient Susan can attest. She doesn't always jump rope religiously, but when she lapses, she reminds herself of how the exercise makes her feel. "When I do get into a good routine of exercise, it replaces that exhilarating feeling or that sense of well-being I get if I drink wine or eat or something," she says. "It replaces whatever that desire or craving is, that thing in the brain. And then it frees me up to think beyond that and look to the future."

THAT WHICH DOESN'T KILL YOU...

It's well known that the way to build muscles is to break them down and let them rest. The same paradigm applies to nerve cells, which have built-in repair and recovery mechanisms activated by mild stress. The great thing about exercise is that it fires up the recovery process in our muscles and our neurons. It leaves our bodies and minds stronger and more resilient, better able to handle future challenges, to think on our feet and adapt more easily.

Regular aerobic activity calms the body, so that it can handle more stress before the serious response involving heart rate and stress hormones kicks in. It raises the trigger point of the physical reaction. In the brain, the mild stress of exercise fortifies the infrastructure of our nerve cells by activating genes to produce certain proteins that protect the cells against damage and disease. So it also raises our neurons' stress threshold.

The cellular stress-and-recovery dynamic takes place on three fronts: oxidation, metabolism, and excitation.

When a nerve cell is called into action, its metabolic machinery switches on like the pilot light in a furnace. As glucose is absorbed into the cell, mitochondria turn it into adenosine triphosphate (ATP)—the main type of fuel a cell can burn—and just as with any energy conversion process, waste by-products are produced. This is oxidative stress. Under normal circumstances, the cell also produces enzymes whose job it is to mop up waste such as free radicals, molecules with a rogue electron that rupture the cell structure while careening around trying to neutralize the electron. These protective enzymes are our internal antioxidants.

Metabolic stress happens when the cells can't produce adequate

?, either because glucose can't get into the cell or because there's not enough of it to go around.

Excitotoxic stress occurs when there is so much glutamate activity that there isn't enough ATP to keep up with the energy demand of the increased information flow. If this continues for too long without recovery, there's a problem. The cell is on a death march—forced to work without food or resources to repair the damage. The dendrites begin to shrink back and eventually cause the cell to die. This is neurodegeneration, the mechanism underlying diseases such as Alzheimer's, Parkinson's, and even aging itself. It's largely through intensive study of these diseases that scientists have discovered the body's natural countermeasures to cellular stress.

And this explains why Mark Mattson, who is chief of the neurosciences lab at the National Institute on Aging, is so stingy with the food for his lab rats. In many of his experiments, Mattson uses dietary restriction to cause mild cellular stress—there isn't enough glucose to produce adequate amounts of ATP—and he's found that mice and rats that are given a third of their normal calories live up to 40 percent longer than average. His work has helped identify protective molecules unleashed during various types of stress, including aerobic exercise.

Some of the most powerful ingredients in the cascade of repair molecules are the growth factors brain-derived neurotrophic factor (BDNF), IGF-1, fibroblast growth factor (FGF-2), and vascular endothelial growth factor (VEGF), which I discussed in chapter 2. BDNF is of particular interest to stress researchers because of its dual role in energy metabolism and synaptic plasticity. It's activated indirectly by glutamate, and it increases the production of antioxidants in the cell as well as protective proteins. And as I've mentioned, it also stimulates LTP and the growth of new nerve cells, strengthening the brain against stress. The advantage of using exercise to inoculate the brain against stress is that it ramps up growth factors more than other stimuli do. In addition to being produced in the brain, FGF-2 and VEGF are also generated by muscle contractions and then travel through the bloodstream into the brain to further support the neurons. This process is a prime example of how the body affects the mind.

The growth factors represent a key link between stress, metabolism, and memory. "The complexity of our brain evolved mainly so we can compete for limited resources," Mattson says. "It makes sense that during evolution, organisms had to compete intellectually to figure out how to find food."

Mattson's latest work will change the way we look at some of our healthiest foods. An enormous industry has sprung up to promote the cancer-fighting properties of foods and products that contain antioxidants. Eat more antioxidant-rich broccoli, the logic goes, and you'll live a longer and healthier life. True, perhaps, but not for the reasons the marketing folks would have you believe.

It turns out that these foods are particularly beneficial not only because they contain antioxidants but also because they contain toxins. "Many of the beneficial chemicals in plants — vegetables and fruits — have evolved as toxins to dissuade insects and other animals from eating them," Mattson explains. "What they're doing is inducing a mild, adaptive stress response in the cells. For example, in broccoli, there's a chemical called sulforaphane, and it clearly activates stress response pathways in cells that upregulate antioxidant enzymes. Broccoli has antioxidants, but at the level you could get from your diet, they're not going to function as antioxidants."

Just as with the nuclear shipyard workers, a mild toxin generates an adaptive stress response that bolsters cells. It's the same process generated by dietary restriction and exercise. The title of one of Mattson's journal articles says it all: "Neuroprotective signaling and the aging brain: Take away my food and let me run."

Resilience is the buildup of these waste-disposing enzymes, neuroprotective factors, and proteins that prevent the naturally programmed death of cells. I like to think of these elements as armies that remain on duty to take on the next stress. The best way to build them up is by bringing mild stress on yourself: using the brain to learn, restricting calories, exercising, and, as Mattson and your mother would remind you, eating your vegetables. All these activities challenge the cells and create waste products that can be just stressful enough. The paradox is that our wonderful ability to adapt and grow doesn't happen without stress—we can't have the good without a bit of the bad.

ENOUGH IS ENOUGH

As with everything in the brain, the stress response depends on a delicate balance of all the ingredients I've mentioned (and many, many that I haven't). If mild stress becomes chronic, the unrelenting cascade of cortisol triggers genetic actions that begin to sever synaptic connections and cause dendrites to atrophy and cells to die; eventually, the hippocampus can end up physically shriveled, like a raisin.

There are a number of scenarios in which the body fails to shut off the flow of stress hormones. The most obvious is simply unrelenting stress. If we never get a break, the recovery process never gets started, the amygdala keeps firing, and the production of cortisol spills over healthy levels. Sometimes the fight-or-flight switch gets stuck in the on position. It can be a function of genetics, according to epidemiological surveys: if you put a random group of people in a stressful public speaking situation, those whose parents suffered from hypertension still show elevated levels of cortisol twenty-four hours after the speech. Or it can be environmental: prenatal rats whose mothers are subjected to repeated stress grow up to have a lower stress threshold than their normal counterparts. Which is to say they get stressed out more easily, both physically and psychologically.

People with low self-esteem also have a lower stress threshold, although scientists aren't sure which condition precedes the other. And anybody, regardless of their nature and upbringing, will exhibit the ill effects of chronic stress if there is no outlet for frustration, no sense of control, no social support. Essentially, if there is no hope, our brains don't shut off the response.

Everybody's threshold for stress is different, and that point can change in response to influences from the environment or our genetics or our behavior or any combination thereof. As with the neurochemistry of the brain, our stress threshold is always changing. While the process of aging naturally lowers the threshold, we can hitch it up quite a few notches through aerobic exercise. There is no specific point at which scientists can say stress shifts from building up to tearing down. But they certainly know the effects when they see them.

THE CORROSIVE EFFECTS OF STRESS

Although stress engraves memories important to survival, too much of it cannibalizes the very structure that does the engraving. While cortisol initially encourages LTP by increasing glutamate transmission in the hippocampus, as well as the flow of BDNF, serotonin, IGF-1, and the like, it activates genes that eventually suppress information reaching those same circuits. One grave context trumps a variety of less important ones. The system becomes less flexible, prioritizing along increasingly rigid lines.

A surplus of glutamate also causes physical damage to the hippocampus. The neurotransmitter acts by allowing electron-snatching calcium ions into the cell, and they create free radicals. Without enough antioxidants on patrol, the free radicals punch holes in the walls, and the cell can rupture and die.

Out at the dendrites, there's trouble too. If the brauches stew too long in the out-of-balance broth of chronic stress, they pull back in an effort to keep the cell from dying, "like a turtle retracting its head," according to McEwen. And because growth factors and serotonin aren't flowing, the process of neurogenesis is interrupted. The new stem cells that are born every day don't turn into new neurons, so there's a shortage of building material to reroute signals and break the cycle.

Monica Starkman at the University of Michigan studies Cushing's syndrome, an endocrine dysfunction in which the body is continually flooded with cortisol. The scientific name for the disorder speaks volumes: hypercortisolism. Its symptoms are eerily similar to those of chronic stress: weight gain around the midsection; breaking down muscle tissue to produce unnecessary glucose and then fat; insulin resistance and possibly diabetes; panic attacks, anxiety, depression, and increased risk of heart disease. One of the many correlations Starkman has shown is that the extent of hippocampal shrinkage and memory loss is directly proportional to elevations in cortisol.

While chronic stress is bullying the hippocampus—pruning its dendrites, killing its neurons, and preventing neurogenesis—the amygdala is having a field day. The stress overload creates more connections in the amygdala, which keeps firing and calling for cortisol, even though there's plenty of the hormone available, and the negative situation feeds on itself. The more the amygdala fires, the stronger it gets. Eventually the amygdala takes control of its partnership with the hippocampus, repressing the context—and thus the connection to reality—and branding the memory with fear. The stress becomes generalized, and the feeling becomes a free-floating sense of fear that morphs into anxiety. It's as if everything is a stressor, and this colors perception and leads to even more stress. "The animal becomes more anxious even while its cognitive skills are being eroded," says McEwen.

When you suffer from chronic stress, you lose the capacity to compare the situation to other memories or to recall that you can grab a jump rope and immediately relieve the stress or that you have friends to talk to or that it's not the end of the world. Positive and realistic thoughts become less accessible, and eventually brain chemistry can shift toward anxiety or depression.

Chronic stress isn't the only cause of anxiety and depression, and it doesn't necessarily lead to either of those disorders. But it is clearly at the root of much of our woe, both physiologically and psychologically, and I'll be returning to the biology of chronic stress throughout the coming chapters.

In a way, the fact that chronic stress underlies many of our problems is great news because we know that how we respond to stress dramatically affects what it does to our bodies and minds. Most of our evolution took place when we were hunter-gatherers, and while there's nothing we can do about that, there is something we can do with that knowledge. As McEwen writes in *The End of Stress as We Know It*, "It's not inevitable or normal for the very system designed to protect us to become a threat in itself."

THE SCIENCE OF BURNING IT OFF

You know by now that the function of the brain is to transmit information, from one synapse to the other, and that this requires energy. Likewise, since exercise influences metabolism, it serves as a powerful way to influence synaptic function, and thus the way we think and feel. Throughout the body, exercise increases blood flow and the availability of glucose, the essentials for cell life. More blood carries more oxygen, which the cells need to convert glucose to ATP and feed themselves. The brain shifts blood flow from the frontal cortex to the middle brain, home to the structures we've talked so much about, the amygdala and hippocampus. This mode of prioritizing might explain why researchers have found that higher cognitive functions are impaired during intense exercise.

It's what happens *after* exercise that optimizes the brain. In addition to raising the fight-or-flight threshold, it kick-starts the cellular recovery process I have described. Exercise increases the efficiency of intercellular energy production, allowing neurons to meet fuel demands without increasing toxic oxidative stress. We do get waste buildup, but we also get the enzymes that chew it up, not to mention a janitorial service that disposes of broken bits of DNA and other by-products of normal cellular use and aging—both of which are thought to help prevent the onset of cancer and neurodegeneration. And while exercise induces the stress response, if the activity level isn't extreme, it shouldn't flood the system with cortisol.

One of the ways exercise optimizes energy usage is by triggering the production of more receptors for insulin. In the body, having more receptors means better use of blood glucose and stronger cells. Best of all, the receptors stay there, which means the newfound efficiency gets built in. If you exercise regularly, and the population of insulin receptors increases if there is a drop in blood sugar or blood flow, the cell will still be able to squeeze enough glucose out of the bloodstream to keep working. Also, exercise increases IGF-1, which helps insulin manage glucose levels.

In the brain, IGF-1 doesn't have as much to do with getting energy into the cells as it does with regulating glucose throughout the body. What's fascinating is that in the hippocampus, IGF-1 increases LTP, neuroplasticity, and neurogenesis. It's another way exercise helps our neurons bind. Exercise also produces FGF-2 and VEGF, which build new capillaries and expand the vascular system in the brain. More and bigger highways means more efficient blood flow.

At the same time, aerobic exercise increases BDNF production. Taken all together, these factors combine forces to make the brain bloom and prevent the damaging effects of chronic stress from taking hold. In addition to cranking up the cellular repair mechanisms, they also keep cortisol in check and increase the levels of our regulatory neurotransmitters serotonin, norepinephrine, and dopamine.

On a mechanical level, exercise relaxes the resting tension of muscle spindles, which breaks the stress-feedback loop to the brain. If the body isn't stressed, the brain figures maybe it can relax too. Over time, regular exercise also increases the efficiency of the cardiovascular system, lowering blood pressure. Cardiologists have recently discovered that a hormone called atrial natriuretic peptide (ANP), which is produced by muscle tissue in the heart, directly tempers the body's stress response by putting the brakes on the HPA axis and quelling noise in the brain. What's so interesting about ANP is that it increases as the heart rate increases during exercise, thus illustrating another pathway by which physical activity relieves both the feeling of stress and the body's response to it.

The stress of exercise is predictable and controllable because you're initiating the action, and these two variables are key to psychology. With exercise, you gain a sense of mastery and self-confidence. As you develop awareness of your own ability to manage stress and not rely on negative coping mechanisms, you increase your ability to "snap out of it," so to speak. You learn to trust that you can deal with it—an extremely important factor for my patient Susan; by jumping rope she inhibits the feeling of stress and the runaway brain activity that can go along with it. "Knowing my brain chemistry—that is the best for me," Susan says. "It's my motivation to get out there. Once I'm in a good place, that motivation is easier—jumping rope almost turns into a need."

Susan has the level of understanding I hope to instill in everyone who reads this book. At every level, from the microcellular to the psychological, exercise not only wards off the ill effects of chronic stress; it can also reverse them. Studies show that if researchers exercise rats that have been chronically stressed, that activity makes the hippocampus grow back to its preshriveled state. The mechanisms by which exercise changes how we think and feel are so much more effective than donuts, medicines, and wine. When you say you feel less stressed out after you go for a swim, or even a fast walk, you *are*.

WHAT PROTECTS THE MIND
PROTECTS THE BODY

Bob was stressed out. It was 1969; he had finished his medical residency; and he was just out of the navy, where he debriefed shell-shocked soldiers sent directly from Vietnam to his base in Boston. But work wasn't the problem—he was a young psychoanalyst and he was quite capable. It was personal stuff: both his father and his father-in-law died in quick succession, and all the emotions he ignored as a teenager after his mother died came rushing back and hit him like a sledgehammer.

Physically too, he was a wreck. He was so stressed that he began having strange choking fits that made it difficult to breathe. He had only recently recovered from a year-long battle with viral meningoencephalitis, a runaway inflammation of the brain that is often fatal, and now he was back in the hospital. This time, he thought it might be throat cancer. There was no indication back then that he would go on to become president of the American Psychoanalytic Association or a faculty member at Harvard or a consultant to Major League Baseball's rookie career development program. In fact, for Dr. Robert Pyles, at age thirty-three, there was no indication he would live another year.

The X-rays revealed a flurry of snowballs in his lungs that turned out to be disseminated sarcoidosis, a cancerlike disease of the lymph system that typically goes on to invade other organs and kills you. "I'm almost positive that these things occurred because I was under a tremendous amount of stress and depression at the time," Pyles says. "I think what happened was, my immune system was so compromised that I got the second disease."

I've discussed the effects of chronic stress on the brain, but the effects on the body are equally powerful. Chronic stress is linked to some of our most deadly diseases. If repeated spikes in blood pressure damage the vessels, plaque can build up at those areas and lead to atherosclerosis. As I mentioned earlier, an unchecked stress response can stockpile fat around the midsection, which studies have shown to be more dangerous than fat stored elsewhere. The overload of cortisol from chronic stress lowers IGF-1 while maintaining glucose levels in the bloodstream, setting up a metabolic imbalance that can lead to diabetes. More broadly, an incessant flow of cortisol clamps down on the immune system, leaving the body wide open to any number of diseases. The results can be deadly.

Pyles wasn't hopeful. At that time, there was no treatment for disseminated sarcoidosis, let alone a cure. One day, he was a Harvard-educated young doctor starting a family and a practice at the dawn of a new decade, and the next, he was handed a death sentence. "I didn't know what to do," he says. "I got more panicked and stressed. What I started doing was, I started to run."

He had been quite an athlete in his school years, but he'd let himself go—to the point where he'd packed 190 pounds onto his five-foot-nine frame. "I had gotten out of college, and like everybody else I didn't do any exercise," he says. "I could only run a maybe a quarter of a mile or half a mile. I would say to myself, If I can run that far, I guess I'm not going to die *today*. After a while I was up to a mile, and then it got to be three miles and then five and then eight. I found that if I got past a certain point, where it was really uncomfortable, it was as though something would click in my psyche, and I could go on for a long time."

Pyles kept running. He wasn't running for his life but for his sanity. The only thing a patient with disseminated sarcoidosis could do was go in for X-rays every three months or so and let them count the snowballs. But for Pyles, the disease seemed to be holding steady.

The months turned into years, the mileage turned into marathons, and then the X-ray film started coming back clearer. After about five years, the disease had disappeared.

This was during an era when if you were sick, a doctor's first recommendation would be to rest. Dr. Kenneth Cooper had only recently coined the term *aerobics*, and we hadn't yet come to accept the health benefits of cardiovascular fitness. Despite his medical training, Pyles didn't recognize that the stress had turned into depression and neither did his analyst. "I think what the running did for me is it gave me some sense of being in control — something I could *do*," Pyles says. "The thing about the depression and the illness is that I felt completely helpless, like I couldn't do anything. There was no way to even fight them at the time."

His doctor wrote up his case for the medical literature, deeming his recovery a miracle cure. But when he suggested it might have something to do with the running, his doctor "just pooh-poohed it completely," Pyles says.

Pyles never intended for running to become such a central part of his life. He gave up smoking a pipe and quit eating meat, because it made him feel heavy. He merged his growing personal interests with his profession, going into practice as a sports psychiatrist for injured athletes who fall into depression because they can't exercise. He's had injuries of his own, of course, but except for a stretch when he was sidelined with a fractured leg, he has run two marathons a year since he started. That's forty-seven in all.

"Back then, doctors had no appreciation whatsoever for exercise as being beneficial in any way," Pyles says. "I still think it's dramatically underappreciated. Particularly in psychiatry. For people who grew up as intellectuals, there's almost an aversion to it."

Pyles attributes this partly to the founding principles of Freudian psychoanalysis. Doing something to avoid talking about our emotions is seen as "acting out." This is the origin of the psychiatrist's couch — the idea is to immobilize the patient and force the emotions to manifest themselves verbally. From this point of view, exercise is a prime example of acting out—dealing with our emotions physically rather than verbally. We're not working on our problems.

Just the opposite turned out to be true for Pyles, who is now seventy-two. His active coping mechanism dramatically redefined both his life and his career. "Exercise saved my life," he says. "I think running really put me back with the unitary nature of body and mind—it's all one thing. We're not split into pieces."

WORK IT OUT

Since the office is a primary source of stress for a lot of people, it's a good place to look for the benefits of exercise. More and more companies are encouraging their employees to take advantage of in-house gyms or health club memberships, and some health insurance companies reimburse clients for club fees. Their generosity is informed by studies showing that exercise reduces stress and makes for more productive employees. In 2004 researchers at Leeds Metropolitan University in England found that workers who used their company's gym were more productive and felt better able to handle their workloads. Most of the 210 participants in the study took an aerobics class at lunchtime, for forty-five minutes to an hour, but others lifted weights or practiced yoga for thirty minutes to an hour. They filled out questionnaires at the end of every workday about how well they interacted with colleagues, managed their time, and met deadlines. Some 65 percent fared better in all three categories on days they exercised. Overall, they felt better about their work and less stressed when they exercised. And they felt less fatigued in the afternoon, despite expending energy at lunchtime.

Other studies show that employees who exercise regularly have fewer sick days. Northern Gas Company employees who participate in the corporate exercise program take 80 percent fewer sick days. General Electric's aircraft division conducted a study during which medical claims by employees who were members of its fitness center went down 27 percent, while nonmembers' claims rose by 17 percent. And according to a report published in the late 1990s by Coca-Cola, health-care claims averaged $500 less for its employees who joined the company's fitness program, compared with those who didn't.

More general research supports the notion that exercise combats stress-related diseases, which, obviously, can keep people out of work. Both stress and inactivity — the twin hallmarks of modern life — play big roles in the development of arthritis, chronic fatigue syndrome, fibromyalgia, and other autoimmune disorders. Reducing stress by any means, and especially exercise, helps patients with their recovery from these diseases. The diseases result from a weakened immune system, and as is evident in the example of Robert Pyles, exercise can dramatically improve immune function. In recent years, doctors have started recommending exercise for cancer patients, both to help boost the immune response and to fend off stress and depression. Nobody says exercise cures cancer, but research suggests that activity is clearly a factor in some forms of the disease: twenty-three of thirty-five studies show an increased risk of breast cancer for those women who are inactive; physically active people have 50 percent less chance of developing colon cancer; and active men over sixty-five have a 70 percent lower chance of developing the advanced, typically fatal form of prostate cancer.

It all comes back to the evolutionary paradox that even though it's much easier to survive in the modern world, we experience more stress. The fact that we're much less active than our ancestors were only exacerbates matters. Just keep in mind that the more stress you have, the more your body needs to move to keep your brain running smoothly.

4

Anxiety

Nothing to Panic About

THE LINE OF questioning began innocuously enough, with the lawyer asking about my background, the books I've written, my areas of expertise. The courtroom was drab and the mood dull, in stark contrast to the underlying drama. Aside from the usual financial considerations, the custody of several children was at stake for the defendant, a patient of mine—I'll call her Amy—who was being divorced by her husband. I had taken the witness stand at the behest of her legal team, to testify about her mental state, and now I was under cross-examination.

Amy is intelligent and attractive, but shy and anxious. She worried all the time, about everything. As her jet-setting husband grew less and less interested in being *her* husband, and his steady stream of criticism grew into a torrent, she began to fear the worst—a repeat of her childhood. The breakup of her family was precisely what she didn't want to happen. When it became clear that divorce was inevitable, she didn't see how she could live with the situation, and in a fit of panic she threatened to kill herself and fled three thousand miles away. Her rash reaction became her legal undoing. The court granted her husband full custody of the children pending the outcome of the trial, restricting her from seeing them except for twice a week. Worse, on the suspicion that she might be unstable, her visits had to be supervised by a court-appointed monitor.

Her husband's lawyer zeroed in on Amy's treatment.

"Is the defendant taking any medication?" she asked, perfectly aware of the answer.

"No, not at the moment," I answered.

"Have you ever prescribed medication for the defendant?"

"Yes. Prozac."

"That's an antidepressant."

"Yes. And it's very effective in treating generalized anxiety disorder."

"And your patient has generalized anxiety disorder?"

"Yes."

"I see. And she's not taking Prozac now. Did you tell her to stop?"

"No. She asked permission, and I told her it was OK." I saw where this was headed: The lawyer was painting Amy as somebody who didn't *want* to get well. In the eyes of the court, treatment means taking medication, so she must not be interested in feeling better. How could someone be trusted to watch over her children if she wouldn't take care of herself?

"But she's been exercising," I interjected. "And she's doing great!"

"Exercise? That's not a proven treatment, is it, doctor?"

"Absolutely. Exercise works a lot like Prozac and our other anti-depressants and antianxiety drugs—"

"That's your opinion," the lawyer interrupted, "but what does it do, exactly?"

"Do you really want to know?" I asked, smiling. "I'm writing a book on the subject."

"Yes, I do."

Perhaps she expected a fuzzy explanation about runner's high. Instead, I cited a few of the clinical trials showing that exercise is as effective as certain medications for treating anxiety and depression. Then I launched into a twenty-minute monologue about what exercise does for the brain, and, specifically, how it had tamed Amy's anxiety and allowed her to master her chaotic feelings in the nine months she'd been my patient. If it was exercise this lawyer wanted to put on trial, I was all for it.

THE CASE

Anxiety is a natural reaction to a threat that happens at a certain point in the stress response, when the sympathetic nervous system and the hypothalamic-pituitary-adrenal (HPA) axis shift into high gear. When you're facing an upcoming speech or a brewing confrontation with your boss, anxiety sharpens your attention so you can meet the challenge. The physical symptoms range from feeling tense, jittery, and short of breath to experiencing a racing heart, sweating, and, in the case of full-blown panic attacks, severe chest pains. Emotionally, what you feel is fear. If you're in a plane that suddenly drops several hundred feet, you and everyone else on board will be edgy and acutely concerned—*are we going to make it?* The nervous system stays alert for a while, hypersensitive to any further turbulence. That's normal.

But if you worry when there's no real threat, to the point where you can't function normally, that's an anxiety disorder. The symptoms crowd your consciousness, your brain loses perspective, and you can't think straight. Clinical anxiety affects about forty million Americans, or 18 percent of the population, in any given year and can manifest in a number of ways. They include generalized anxiety disorder, panic disorder, specific phobias, and social anxiety disorder. They all share the physical symptoms of the severe stress response as well as a similar dysfunction in the brain, namely a cognitive misinterpretation of the situation. The common denominator is irrational dread. The differences are mostly a matter of context.

Someone with generalized anxiety disorder tends to respond to normal situations as if they were threatening—the Nervous Nelly who is afraid of her own shadow or the worrier who sees stressors everywhere. People who suffer from panic disorder seem perfectly at ease most of the time but then are blindsided by crippling fear and physical pain that can be mistaken for a heart attack. Panic is the most intense form of anxiety, and it is at the root of all phobia—a paralyzing fear of a specific object or situation that instills a powerful and often unreasonable compulsion to avoid the source (spiders for the arachnophobe, open spaces for the agoraphobe). Probably the most common phobia is social anxiety disorder, which I think of as performance anxiety in everyday interactions. Most of us experience social anxiety at some point or in certain situations, but the disorder is more intense than just being shy on occasion. It's a consuming fear of any social situation that might involve meeting or talking to people or even being seen by others, and it's more common than most people realize, afflicting fifteen million Americans. Social anxiety disorder takes a serious toll on quality of life.

All of these forms of anxiety can bleed into and feed off one another, and they often seed other disorders such as depression. It's possible to have panic disorder without having generalized anxiety disorder and vice versa, but often panic disorder turns into generalized anxiety due to fear of the next attack. Some people also have anxiety sensitivity, which complicates any form of the disorder. Your heart rate or breathing might increase for an unrelated reason, and when you sense this physical arousal, that awareness alone can trigger a state of anxiety or panic. You lose control because you feel like you're going to lose control. If you begin to fear the fear—whether it's mental or physical—anxiety can quickly spiral out of control.

Amy is a textbook case of generalized anxiety disorder, with shades of panic disorder and social anxiety disorder. She displays both the *state*—hyperalert, tense, expecting the worst—and the *trait*, which is the deeper, more ingrained tendency to slip into the state. All her life, she's had anxiety sensitivity, and as her marriage crumbled it only intensified. She began responding to every stressor, whether or not it was truly threatening, as if it were a matter of survival, overreacting and doing a lot of damage to herself and her relationships in the process.

You couldn't ask for a more anxiety-prone situation than the one Amy fell into. Her husband held de facto control over her time with the kids; she had to visit a psychologist who reported back to the court; and everyone in town knew what was going on. Her social anxiety was in full bloom during her supervised visits—she essentially had to perform for the court-appointed monitor—and she was afraid that she would slip up and somehow give her husband more legal ammunition. She was being judged on her mental health, and the more she worried about how she came off, the worse her symptoms became. In this environment, Amy began to doubt her own abilities as a mother, even though she'd been perfectly competent. She desperately wanted to rescue herself and get her children back, but she was in no condition for a fight, feeling like a nervous wreck with no control over her anxiety. It was a sickening spiral: constantly on the verge of panic, she felt like she couldn't stand up for herself or accomplish anything.

When we are in this state, we begin to anticipate that everything is going to be a disaster, and so we try to avoid everything, and our world begins to shrink. Amy had retreated inside her new apartment since her marriage derailed and had become completely withdrawn from her friends and family.

THE DEFENSE

Unlike the lawyer's portrayal, Amy was keenly interested in getting better. It's not criminal or even unusual that she didn't like the idea of taking medicine, but she tried Prozac for a stretch anyway.

Although it calmed her nerves, it left her feeling unmotivated, and she stopped. She had been practicing Kripalu yoga, which also calmed her down, but still she suffered, so I encouraged her to add in aerobic exercise. She bought an elliptical trainer for her apartment, by far a more palatable option than stepping outside her safety zone.

Gradually, she got into a routine, logging thirty minutes every morning. She had very little she could enjoy during this period, but she started having fun with the exercise. She described how she incorporated upper-body twists while pumping away on the trainer, and she followed her aerobic session with an hour of yoga (which has been proven to reduce anxiousness). She was gaining a sense of control over her state of anxiety, a vital step toward conquering the trait. She quickly learned that if she became anxious or panicky at home, she could jump on the elliptical for ten or fifteen minutes to quell the feelings on the spot (in the same way my patient Susan used her jump rope to cope with stress).

Amy rediscovered her motivation through the movement. Not only did she stop worrying all the time, but she also began to see herself as being active rather than passive. She didn't feel frozen anymore, and she reengaged in other areas of her life. She got back to her hobbies and friends, which allowed her to reconnect with the good things about herself. Now she doesn't feel like a rat in a corner, cowering or startling at every disturbance. A casual observer might say that Amy has come out of her shell, but the ripple effects of exercise on her personality are much more profound. She carries herself as if she were on solid ground.

The truth is, her situation hasn't changed all that much — just her response to it and thus her attitude. She says she uses exercise the same way someone else might take a shot of whiskey or alprazolam (Xanax) to calm her nerves. Her strategy has noticeably lowered her anxiety sensitivity, which allows the brain to learn its way out of the trap.

THE EVIDENCE

In 2004 a researcher named Joshua Broman-Fulks from the University of Southern Mississippi tested whether exercise would reduce anxiety sensitivity. He found fifty-four college students with generalized anxiety disorder who had elevated anxiety sensitivity scores and who exercised less than once a week. He randomly divided his sedentary subjects into two groups, both of which were assigned six twenty-minute exercise sessions over two weeks. The first group ran on treadmills at an intensity level of 60 to 90 percent of their maximum heart rates. The second group walked on treadmills at a pace of one mile per hour, roughly equal to 50 percent of their maximum heart rates.

Both regimens tended to reduce anxiety sensitivity, but rigorous exercise worked more quickly and effectively. Only the high-intensity group felt less afraid of the physical symptoms of anxiety, and this distinction started to show up after just the second exercise session. The theory is that when we increase our heart rate and breathing in the context of exercise, we learn that these physical signs don't necessarily lead to an anxiety attack. We become more comfortable with the feeling of our body being aroused, and we don't automatically assume that the arousal is noxious.

This is a key finding given the notion of anxiety as a cognitive misinterpretation. By using exercise to combat the symptoms of anxiety, you can treat the state, and as your level of fitness improves, you chip away at the trait. Over time, you teach the brain that the symptoms don't always spell doom and that you can survive; you're reprogramming the cognitive misinterpretation.

The fact that aerobic exercise works immediately to fend off the state of anxiety has been well established for many, many years. It's only more recently, however, that researchers have started to pin down how this works.

In the body, physical activity lowers the resting tension of the muscles and thus interrupts the anxiety feedback loop to the brain. If the body is calm, the brain is less prone to worry. Exercise also produces calming chemical changes. As our muscles begin working, the body breaks down fat molecules to fuel them, liberating fatty acids in the bloodstream. These free fatty acids compete with tryptophan, one of the eight essential amino acids, for slots on transport proteins, increasing its concentration in the bloodstream. The tryptophan pushes through the blood-brain barrier to equalize its levels, and once inside, it's immediately put to use as the building block for our old friend serotonin. In addition to the boost from tryptophan, the higher brain-derived neurotrophic factor (BDNF) levels that come along with exercise also ramp up levels of serotonin, which calms us down and enhances our sense of safety.

Moving the body also triggers the release of gammaaminobutyric acid (GABA), which is the brain's major inhibitory neurotransmitter (and the primary target for most of our antianxiety medicines). Having normal levels of GABA is crucial to stopping, at the cellular level, the self-fulfilling prophecy of anxiety—it interrupts the obsessive feedback loop within the brain. And when the heart starts beating hard, its muscle cells produce a molecule called atrial natriuretic peptide (ANP) that puts the brakes on the hyperaroused state. ANP is another tool the body uses to regulate the stress response, which I'll explain more later.

As for the trait, the majority of studies show that aerobic exercise significantly alleviates symptoms of any anxiety disorder. But exercise also helps the average person reduce normal feelings of anxiousness. One interesting study in 2005 measured the physical and mental effects of exercise in a group of Chilean high school students for nine months. The researchers divided 198 fifteen-year-olds into two groups: the control group continued with a once-a-week, ninety-minute gym class, and the other embarked on a program of its own design, rigorously exercising during three ninety-minute sessions per week throughout the school year. The study was meant to assess general mood changes in a healthy population, but scores relating to anxiousness really stood out on the students' psychological tests. The experimental group's anxiety scores dropped 14 percent versus a statistically insignificant 3 percent for the control group (an improvement that could be explained by the placebo effect). Not coincidentally, the experimental group's fitness levels improved 8.5 percent versus 1.8 percent for the control group. Clearly, there is a connection between how much you exercise and how anxious you feel.

FEAR ITSELF

Anxiety is fear, but what is fear? In neurological terms, fear is the memory of danger. If we suffer from an anxiety disorder, the brain constantly replays that memory, forcing us to live in that fear. It all starts when the amygdala sounds the survival call, but unlike the normal stress response, in anxiety the all-clear signal isn't working properly. Our cognitive processors fail to tell us there is no problem or that it has passed and we can relax. There is so much noise in the mind from the sensory input of physical and mental tension that it clouds our ability to clearly assess the situation.

The misinterpretation stems partly from an amygdala that isn't effectively controlled by the prefrontal cortex. One of the correlations scientists have found among people with generalized anxiety disorder is brain scans that show the area of the prefrontal cortex responsible for sending cease-and-desist signals to the amygdala as being smaller than it should be. Left unchecked, the overexcited amygdala tags too many situations as challenges to survival and burns them into memory. The fear memories form connections with each other, and the anxiety snowballs. Eventually, the amygdala overwhelms attempts by the hippocampus to tone down the fight-or-flight response by putting the fear in context. As the snowball grows and more and more memories become associated with fear, your world shrinks.

A patient of mine who suffers from social anxiety disorder is a good example of how the fear can snowball, as well as how we can rein it in. She's a late-twenties office manager who dreaded social gatherings, meeting new people, and even small talk with someone she already knew. Ellen, I'll call her, would get butterflies and a dry mouth just thinking about going to a cocktail party, and once there, she couldn't wait for the first drink to hit. Like most people with social anxiety, she felt on display, terrified that she would do something embarrassing or humiliating. Afterward, she would go home and berate herself for her "performance."

All this made it extremely stressful for Ellen to manage her seven employees. She wished she could stop apologizing for assigning tasks, but her anxiety kept her from acting like a boss. It wasn't right, she knew, to plead with people to do their jobs, but she felt terribly guilty for asking anything of them and then worried that she was asking too much. As her sense of authority eroded, she only became more anxious and began avoiding contact with anyone in the office for fear someone would spot her weaknesses.

What makes anxiety so tricky to treat is that survival-related memories trump existing memories. Say you walk by a certain house every night on your way home from work, and then one night a dog comes rushing out and attacks you. From that moment on, you'll skirt that house, because the memory of the attack stands out against all the times you passed by safely. Even if a fence is built and you're the most logically minded person on earth, you'll still feel a little jumpy walking by. Once the fear memory is wired in, that particular circuit stays there. Which is to say, fear is forever.

Contrary to what scientists originally assumed, studies comparing MRI scans of brain activity in adults with and without anxiety disorders show no difference in how the amygdala responds to a legitimately frightening stimulus (such as pictures of fear-stricken faces, which have a powerful effect because human beings are programmed to interpret facial expressions as survival cues).

The difference is in how they respond to a *nonthreatening* stimulus. Whereas most people will show a sharp decrease in amygdala activity when presented with a benign picture, those with anxiety disorder have almost the same activity level as if they were confronting fear—they cannot discriminate between danger and safety. Research psychiatrist Daniel Pine, who is the chief of the section on development and affective neuroscience at the National Institutes of Mental Health, sees it this way: "Patients with anxiety disorders have a learning deficit."

There may be genetic factors underlying the dysfunction of learning circuits in anxiety. Researchers recently studied a gene variation that prevents BDNF from fostering nerve connections, which results in impaired memory. In the experiment, mice with the mutant BDNF gene that were put in an anxiety-provoking situation didn't get the relief they should have from Prozac. The antidepressant worked fine for normal mice under the same circumstances. This suggests that BDNF might be an essential ingredient in combating anxiety, probably because it helps wire in positive memories that create a detour around the fear.

I think this is a big reason why exercise is so effective at treating not just the state of anxiety—by relieving muscle tension and increasing serotonin and GABA—but also the trait of anxiety. Exercise gives neurons everything they need to connect, and if we direct that process we can have a huge impact on teaching the brain to cope with the fear.

My patient Ellen came to me on the usual selective serotonin reuptake inhibitor (SSRI) antidepressant, and while the medicine helped, it didn't solve the underlying problem. Naturally, I talked to her about exercising. She acknowledged that she felt less anxious after running but said she was too busy to make it a priority. I told her the irony was that she'd feel less harried if she took the extra time to exercise, and after a bit of goading (and fine-tuning her medication), she began going to the gym before work. It quickly became clear that on the days she missed her workout, she felt more flustered and less willing to interact with anyone at work, including new clients. Then she shifted into high gear and made a point of going every morning. If she missed her favorite aerobic class, she would run on the treadmill for twenty minutes, and she's been at it for a year or so.

Now Ellen feels like she's able to be more assertive and straightforward with her employees, and the more she interacts with them, the bolder she becomes. A huge part of the problem with social anxiety, whether it's at the level of Ellen's phobia or milder social apprehension, is that the more we withdraw, the less practice we get interacting, and the scarier the prospect becomes. It might sound silly that someone would need to practice what comes naturally to many people, but it's not silly at all. This is the genius of Paul Zientarski's freshmen square dance class at Naperville Central High School—all the kids practice small talk in the same situation, incrementally over the course of the semester, neutralizing any fear they might have. For Ellen, exercise was a tool that calmed her nerves enough for her to test the waters. Just as anxiety can feed on itself, so can courage.

THE PAIN OF PANIC

Panic is the most painful form of anxiety, and it illustrates in the extreme how paralyzing any of this family of disorders can be. When I first came across a case of panic disorder, I was shocked at how debilitating it was. I was a third-year psychiatric resident at Massachusetts Mental Health Center and was also seeing patients at outlying social services offices. A woman had been dragged in by her husband because she was depressed and all but refused to leave the house. She had been to the emergency room on more than one occasion for what felt like a heart attack, and she described in vivid detail how she had been certain she was dying. Each time, the doctor reported that her heart was fine, and she started to wonder if she was crazy.

Panic doesn't cause heart failure, but it sure feels that way. Muscle tension and hyperventilation cause severe chest pains. Then, because the rapid, shallow breathing expels too much carbon dioxide, the blood's pH level drops, triggering an alarm from the brain stem that causes muscles to constrict even more. (This is why breathing into a paper bag stops us from hyperventilating: it forces us to rebreathe the carbon dioxide.)

Living with panic means avoiding anything that might set off another frightening episode. You withdraw into an emotional fetal position, and the fear leads to a desperate need for control — whatever is necessary to maintain a stable and safe environment. This manifests itself in various ways: passive-aggressiveness, which is one way of trying to control others; compulsiveness, to keep the fear triggers at bay; and overall inflexibility. My patient knew something was wrong, but the panic had taken over to such an extent that the constellation of symptoms warped the picture of the real problem.

The major treatment for anxiety and depression at the time, in the late 1970s, was psychotherapy. We just didn't use drugs much. But the field was beginning to shift toward a biological interpretation of mental health, and studies were cropping up about treating anxiety with imipramine, a tricyclic antidepressant that had been around for twenty years. It manipulates the interaction of norepinephrine and serotonin in a part of the brain stem known as the locus coeruleus, which regulates basic life functions such as breathing, waking, heart rate, and blood pressure. As such, this area monitors blood pH levels and is the origin of the alarm signals that trigger the amygdala in a panic attack. The drug stabilizes the arousal system so the panic button isn't so easily tripped. It worked almost immediately for my patient, and over days and weeks in the absence of anxiousness she slowly let down her guard.

By controlling the fear, we were able to move forward in therapy. Imipramine gave her back her freedom.

Another class of drug that became popular for treating various forms of anxiety around the same time was beta-blockers, which calm down the sympathetic nervous system. They block epinephrine receptors in the brain and the body and thus prevent epinephrine from elevating blood pressure, heart rate, and breathing during times of stress or anxiety. Often used for heart patients to reduce blood pressure, beta-blockers break the anxiety feedback loop to the brain that otherwise keeps the amygdala on alert. In quelling the bodily symptoms of anxiety, beta-blockers diffuse panic attacks before they explode. They're also useful for people with social anxiety or stage fright. It's exceedingly common for classical musicians to take beta-blockers before performances because it prevents them from sweating and tensing up, which can really interfere with their ability to play. (It must be hard to play a trombone with stiff lips!)

Sometimes people with panic disorder are treated with both imipramine and beta-blockers—the first to quash the fear and the second to relax the body. The real point of understanding how these drugs work is that they provide an explanation for how exercise works. As it turns out, exercise impacts the same pathways as these medications—it puts a safety on both triggers.

PUSHING THROUGH THE PAIN

For several decades, common medical wisdom suggested that patients suffering from panic should *avoid* exercise. It could be dangerous! Or so we thought, based on research from the late 1960s. Some patients reported that the physical manifestations of exercise—increased heart rate, blood pressure, rapid breathing—magnified their fear, presumably because they felt just like the symptoms of anxiety. It turned out that some people with anxiety disorders had elevated levels of lactic acid in their blood compared to nonanxious exercisers, and researchers found that infusing anxiety patients with lactic acid induced panic attacks. Doctors began advising patients with any form of anxiety to avoid exercise, lest it trigger an attack. Better to stay still.

This logic persisted despite a number of follow-up studies that disproved the hypothesis. Although the medical literature tells us that a handful of patients did panic during exercise, the vast majority showed just the opposite effect. In fact, 104 studies on exercise and anxiety reported between 1960 and 1989, showed that exercise alleviates anxiety, but most of them didn't meet the randomized, double-blind, placebo-controlled trial criteria necessary for scientists to count them as medical fact. While analytical researchers might say there isn't sufficient data to prove that exercise reduces anxiety, others will tell you that they don't bother studying the issue because it's just common sense.

Consequently, the first randomized, placebo-controlled study to compare exercise to drugs in treating clinically diagnosed panic disorder was undertaken in 1997. German psychiatrist Andreas Broocks conducted a ten-week trial in which he divided forty-six patients with at least mild panic disorder into three groups: regular exercise, a daily dose of clomipramine (a close relative of imipramine), or a daily placebo pill.

All three treatments improved symptoms in the first two weeks, including the placebo! The clomipramine had the quickest and most dramatic effect, immediately and steadily relieving symptoms. In the exercise group, after initial relief anxiety scores leveled out somewhat until the last four weeks, when they rapidly decreased. (The placebo group experienced a return of symptoms as the trial progressed.) At the end of ten weeks, the clomipramine group and the exercise group ended up at the same level of improvement on a variety of tests. Both groups were in remission.

Why did exercise take longer? According to another scientifically rigorous study, by Andreas Ströhle in 2005, it shouldn't have.

Ströhle showed that thirty minutes of treadmill running significantly reduces panic attacks as compared with quiet rest (by a ratio of two to one), indicating that its effect can be immediate in some cases. The lag in exercise-induced relief in Broocks's study probably had to do with the way it was set up. All but one person in the exercise group had agoraphobia, some of them quite severely, and others regarded exercise as flat-out "dangerous," which means they believed the very act of walking or running outside would be difficult. They had to confront their fear to follow the study's instructions. You can't simply tell someone with agoraphobia to go for a four-mile run and everything will be OK, so Broocks had them ease into the regimen. They were asked to find a four-mile route near their home and merely complete it three or four times a week—walking at first if necessary. They were then encouraged to include short stints of running and gradually increase them in length; they weren't expected to run the entire route until after the sixth week. Two of the patients actually had panic attacks while running but kept going and they subsided.

In the Broocks experiment, everyone in the clomipramine group stuck with the therapy throughout the study, despite significant side effects, including dry mouth, sweating, dizziness, tremors, erectile dysfunction, and nausea. The exercise group, which along with the placebo group had several patients drop out, reported minor side effects of the sort expected for someone starting up a new exercise routine, including temporary muscle and joint discomfort.

In a six-month follow-up, the exercise patients who were the most fit had the lowest anxiety scores. In the end, the exercise group landed in the same healthy place as the clomipramine group, and they did so of their own accord. There's certainly nothing wrong with taking medicine, but if you can achieve the same results through exercise, you build confidence in your own ability to cope. This is a significant advantage not just for patients with full-blown anxiety disorders, but for anyone. We all face situations in everyday life that cause fear and anxiousness. The trick, as my patient Amy illustrates, is in how you respond.

THE MISSING CONNECTION

The mind-set that any sound treatment for anxiety must involve medicine isn't restricted to the courtrooms of divorce proceedings. In 2004 the *New England Journal of Medicine* (NEJM) published a review of treatments for generalized anxiety disorder that failed to even mention exercise. It was primarily a rundown of our most common antianxiety drugs, with a nod to therapy and relaxation. Of the thirteen pharmaceuticals charted in the review, all bear a formidable list of possible side effects. None have been endorsed by the U.S. Food and Drug Administration as explicitly safe during pregnancy—not an incidental point given that women are twice as likely to suffer from anxiety and depression as men.

The article was positioned as advice for doctors, but how is it that a summary of treatments for general anxiety disorder in the bible of medical research simply left out exercise? It's a case of what I would call clinical blindness. The mounting research on the neurological and psychological benefits of exercise seems to be hidden in plain sight.

Interestingly, it was the cardiologists who spoke up. The *NEJM* published a letter from doctors Carl Lavie and Richard Milani of the Ochsner Clinic Foundation in New Orleans. It read, in part, that the author "discusses generalized anxiety disorder and its treatment with pharmacologic agents and psychotherapy. We are surprised, however, that there is no mention of exercise as an additional means of treating anxiety." The letter noted that cardiologists are interested in anxiety as a risk factor for heart problems, and then pointed out, "Exercise training has been shown to lead to reductions of more than 50 percent in the prevalence of the symptoms of anxiety. This supports exercise training as an additional method to reduce chronic anxiety."

The letter was a polite way of saying that the original article missed the boat. Lavie has written more than seventy papers on exercise and the heart, eleven of which focus on anxiety. Every single one of his studies has shown a marked improvement in anxiety and depression.

The importance of this exchange is that it's a case of cardiologists ("real" doctors) taking psychiatrists to task about how to treat the whole patient. If we go all the way back to Hippocrates, the wisdom of the day was that emotions come from the heart and that that's where treatment should start for maladies of mood. Modern medicine has separated mind and body, but it turns out that, in a very concrete way, Hippocrates had it right from the start. Just in the past ten years scientists have begun to understand how a molecule that originates in the heart plays on our emotions.

ANP is secreted by heart muscles when we exercise, and it makes its way through the blood-brain barrier. Once inside, it attaches to receptors in the hypothalamus to modulate HPA axis activity. (ANP is also produced directly in the brain, by neurons in the locus coeruleus and in the amygdala—both key players in stress and anxiety.) ANP has been shown in both animal and human studies to have a calming effect, and researchers suspect it to be a major link between exercise and anxiety. In 2001 one of the first studies to verify the role of ANP in anxiety compared patients with panic disorder to those without. They were randomly assigned an injection of ANP or placebo, and then received a dose of an abdominal hormone called cholecystokinin tetrapeptide (CCK-4), which induces anxiety and panic. ANP significantly reduced panic attacks in both groups, while the placebo did not.

During a panic attack, there is a surge of corticotropin-releasing factor (CRF), which induces anxiety in its own right and also floods the nervous system with cortisol. ANP seems to work against CRF's efforts to put us into a frenzy, like a drag brake on the HPA axis. And studies in women show that levels of ANP triple during pregnancy, suggesting a built-in survival strategy to protect the baby's developing brain from the potentially toxic effects of stress and anxiety.

In one study of patients with severe heart failure, those with the highest ANP levels had the lowest levels of anxiety. None of them had anxiety disorders, but doctors were interested in their anxiety because it has a major influence on how well heart surgery patients recover. ANP directly dampens the sympathetic nervous system's response by stemming the flow of epinephrine and lowering the heart rate, and it also seems to reduce the *feeling* of anxiousness, which is paramount. And we know that among panic disorder patients, those who have frequent attacks have a deficit of ANP in their bloodstreams.

In 2006 a group of neuropsychiatrists from Berlin led by Andreas Ströhle looked at whether ANP was a critical element in the calming effect of aerobic exercise. For ten healthy patients who agreed to have panic-inducing injections of CCK-4, walking for thirty minutes on a treadmill (at a moderate pace) significantly increased concentrations of ANP while simultaneously lowering feelings of anxiety and panic. Ströhle pointed out that correlations don't equal causality, but he wrote, "ANP may be a physiologically relevant link between the heart and anxiety-related behavior."

FACING THE FEAR

If fear is forever, how can we hope to snuff out anxiety? The answer lies in a neurological process called fear extinction. While we can't erase the original fear memory, we can essentially drown it out by creating a new memory and reinforcing it. By building up parallel circuitry to the fear memory, the brain creates a neutral alternative to the expected anxiety, learning that everything is OK. By wiring in the correct interpretation, the trigger is disconnected from the typical response, weakening the association between, say, seeing a spider and experiencing terror and a racing heart. Scientists call it reattribution.

We can force the brain to trade fear memories for neutral or positive ones through a form of psychology called cognitive behavioral therapy (CBT). Studies show that CBT is about as effective as SSRIs in treating anxiety, although varied results suggest that the quality of the therapy is crucial. The strategy is to expose the patient to the source of fear in small doses in the company of the therapist. When we experience the symptoms without the panic, the brain goes through a cognitive restructuring. We build connections in the prefrontal cortex that help calm the amygdala, which makes us feel safe, and then the brain records a memory of that feeling. When we add in exercise, we get the neurotransmitters and neurotrophic factors bolstering the circuits between the prefrontal cortex and the amygdala, providing further control and creating a positive snowball effect.

Psychologist and distance runner Keith Johnsgard found that conducting CBT in the context of exercise has particularly powerful results. In his book *Conquering Depression and Anxiety through Exercise,* he explains how he uses running as a mode of cognitive restructuring to treat agoraphobia. After several rapport-building sessions, he accompanies patients to an empty mall parking lot in the early morning and has them do a series of sprints. Nobody else is around, and they feel safe in his presence. He has already determined how far they can sprint before coming to exhaustion, and—this is the clever part—Johnsgard marks off that distance from the main door of the mall, and then has them sprint from his side *toward* the mall. The idea is that they reach the height of their fear in a state of full physical arousal, without the panic. If they feel a panic attack coming on, they are instructed to stop, turn, and walk back to him. They run toward fear and walk toward safety.

Eventually they should overcome the anxiety of entering the mall and be able to venture inside for increasingly longer stretches. He says that often he sees improvement after only a half dozen sessions. "In essence," he writes, his approach is a matter of "getting back on the horse that threw you." Teaching the brain that we can survive is crucial to overcoming the anxiety.

This approach fits into a broader concept highlighted by New York University neuroscientist Joseph LeDoux, a renowned fear expert. Shortly after the terrorist attacks of September 11, 2001, LeDoux and coauthor Jack Gorman published an article in the *American Journal of Psychiatry* titled, "A Call to Action: Overcoming Anxiety through Active Coping." Essentially, active coping means doing something in response to whatever danger or problem is causing anxiety rather than passively worrying about it. It doesn't specifically imply physical activity, but certainly exercise qualifies as a mode of active coping. And as it turns out, movement may not be an incidental aspect of active coping.

LeDoux discusses how, by making a decision to act in the face of anxiety, we literally shift the flow of information in the brain, forging new pathways. An area of the amygdala called the central nucleus is responsible for creating the negative snowball effect—linking nonthreatening stimuli with legitimately threatening stimuli. The resulting fear memory is the connection between the trigger and the anxiety.

LeDoux has shown in rats that the signals can be redirected so that instead of pulsing through the central nucleus of the amygdala, they go through the basal nucleus, which connects to the body's motor circuits. If the same is true in humans, simply by taking action we're circumventing the mechanism for the fear memory. The basal nucleus is the action pathway, and we can even spark it with thought. For one of my patients, who was traumatized by losing both his job and his girlfriend, I suggested that he start each day by getting to the gym, to keep from stewing in the trauma. He could also shift the flow from fear to action circuits by making a list of potential employers to call—a more classic example of active coping—but it wouldn't affect the brain as broadly. By doing something other than sitting and worrying, we reroute our thought process around the passive-response center and dilute the fear, while at the same time optimizing the brain to learn a new scenario. Everyone's initial instinct in the face of anxiety is to avoid the situation, like a rat that freezes in its cage. But doing just the opposite, we engage in cognitive restructuring, using our bodies to cure our brains.

OUTRUNNING THE FEAR

The elegance of exercise as a way to deal with anxiety, in everyday life and in the form of a disorder, is that it works on both the body and the brain. Here's how:

1. It provides distraction. Quite literally, moving puts your mind on something else, just as using the elliptical trainer helped my patient Amy break out of her acute state of anxiety and focus on something other than the fear of her next panic attack. Studies have shown that anxious people respond well to any directed distraction—quietly sitting, meditating, eating lunch with a group, reading a magazine. But the antianxiety effects of exercise last longer and carry the other side benefits listed here.

2. It reduces muscle tension. Exercise serves as a circuit breaker just like beta-blockers, interrupting the negative feedback loop from the body to the brain that heightens anxiety. Back in 1982 a researcher named Herbert de Vries conducted a study showing that people with anxiety have overactive electrical patterns in their muscle spindles and that exercise reduced that tension (just as beta-blockers do). He called it the "tranquilizing effects of exercise." Reducing muscle tension, he found, reduced the feeling of anxiety, which, as I've explained, is important to extinguishing not just the state but the trait of anxiety.

3. It builds brain resources. You know by now that exercise increases serotonin and norepinephrine both in the moment and over the long term. Serotonin works at nearly every junction of the anxiety circuitry, regulating signals at the brain stem, improving the performance of the prefrontal cortex to inhibit the fear, and calming down the amygdala itself. Norepinephrine is the arousal neurotransmitter, so modulating its activity is critical to breaking the anxiety cycle. Physical activity also increases the inhibitory neurotransmitter GABA as well as BDNF, which is important for cementing alternative memories.

4. It teaches a different outcome. One aspect of anxiety that makes it so different from other disorders is the physical symptoms. Because anxiety brings the sympathetic nervous system into play, when you sense your heart rate and breathing picking up, that awareness can trigger anxiety or a panic attack. But those same symptoms are inherent to aerobic exercise — and that's a good thing. If you begin to associate the physical symptoms of anxiety with something positive, something that you initiated and can control, the fear memory fades in contrast to the fresh one taking shape. Think of it as a biological bait and switch — your mind is expecting a panic attack, but instead it ends up with a positive association with the symptoms.

5. It reroutes your circuits. By activating the sympathetic nervous system through exercise, you break free from the trap of passively waiting and worrying, and thus prevent the amygdala from running wild and reinforcing the danger-filled view of what life is presenting. Instead, when you respond with action, you send information down a different pathway of the amygdala, paving a safe detour and wearing in a good groove. You're improving alternate connections, actively learning an alternative reality.

6. It improves resilience. You learn that you can be effective in controlling anxiety without letting it turn into panic. The psychological term is self-mastery, and developing it is a powerful prophylactic against anxiety sensitivity and against depression, which can develop from anxiety. In consciously making the decision to do something for yourself, you begin to realize that you *can* do something for yourself. It's a very useful tautology.

7. It sets you free. Researchers immobilize rats in order to study stress. In people too, if you're locked down—literally or figuratively—you'll feel more anxious. People who are anxious tend to immobilize themselves—balling up in a fetal position or just finding a safe spot to hide from the world. Agoraphobics feel trapped in their homes, but in a sense any form of anxiety feels like a trap. The opposite of that, and the treatment, is taking action, going out and exploring, moving through the environment. Exercising.

PLAN A COUNTERATTACK

One big difference between combining exercise with antianxiety medicine and using medicine alone is that while drugs like benzodiazepine—and alcohol for people who are self-medicating—quickly stifle anxiety, they don't guarantee that you'll learn a different response to the fear. People with anxiety often have a hard time knowing or choosing what they want from life. In fact, all that most chronically anxious people want is not to be anxious. Activity or exercise can help them move *toward* something.

I don't look at exercise and medicine as an either-or proposition. Exercise is another tool at your disposal, and it's handy because it's something you can prescribe for yourself, whether you have a definable disorder or just feel anxious at times. Like most people do. And I'm certainly not a pharmacological Calvinist—I don't tell people to pull themselves up by their bootstraps or that it's a sin or weakness to rely on medicine.

I recently took on a patient with panic disorder who is a high school senior. He had his first episode when he was six, which implies that he has a predisposition to it, and of late it has been getting worse because of the pressures of preparing for college. Whenever he goes for a run, as soon as his heart starts pumping he feels anxious about the possibility of panicking, and he worries that he'll drop dead of a heart attack and that nobody will find him. Sometimes he'll stop and just start crying. But he also knows, rationally, that if he pushes through his sensitivity to the physical arousal, the feeling subsides. Would I recommend that he quit taking Zoloft? Absolutely not.

First of all, he's deeply phobic about a panic attack coming on. And because panic disorder is so frightening, I usually start with medicine. Taking a pill doesn't require a lot of effort, and in some cases it works like flipping a switch to diffuse the trigger. But, as I mentioned, it doesn't necessarily lead to a permanent change, and the relearning process needs to happen for long-term relief. Why not come out with both guns blazing? I think combining medicine with exercise can be a great approach. Medicine provides immediate safety, and exercise gets at the fundamentals of anxiety.

This is particularly important with respect to children, because kids who have anxiety are more likely than their peers to develop depression later in life. One long-term study followed seven hundred children into adulthood. Of those who suffered from anxiety as children, most grew out of it. But of those who developed a mood disorder, in two-thirds of the cases the problem started as preadolescent anxiety. What's tragic here is that anxiety is relatively easy to treat, but it often goes undiagnosed in children — the anxious kids are sitting quietly at the back of the class, terrified. Nobody notices there is something wrong because these kids are well behaved. Meanwhile, the anxiety is wearing in negative patterns in their brains that can become entrenched and set these kids up for future problems.

I told the young man that the A-number-one thing he needed to do was exercise *with* somebody. That goes for anyone who is panicky. It offers a sense of safety, but it also increases levels of serotonin immediately, just from being around another person. In his case, I suggested that he exercise at home, or close to it, until he started to associate his elevated heart rate with a positive experience. He needed to find a type of exercise he enjoyed, and, because his panic seemed to have a strong genetic component, I told him he'd really have to work at it. He'd have to start doing at least fifteen minutes of rigorous aerobic exercise a day—running, swimming, biking, rowing, or whatever would get his heart pounding. Intensity was particularly important in his case because of the evidence that only rigorous exercise alleviates sensitivity to the physical arousal of anxiety.

Like almost every adolescent I've seen, this young man didn't want to be on medicine. He asked if he could stop taking it, and I told him that over time, if he got into the exercise and maybe CBT, he should be able to lower his anxiety sensitivity. Eventually, I said, my guess was that he'd be able to reduce his medication and maybe eliminate it altogether. But nobody really knows whether exercise can entirely replace medication. Our brains are just too complex.

A lot of people who are treated for panic disorder can go on to have a completely different kind of life. The farther they get from their last panic episode, the less likely they are to have another panic episode. The same holds true for any brand and any degree of anxiety. The more your life changes, the more you engage with the world, the more likely you are to put the anxiety behind you for good. Exercise can have an even more dramatic effect on milder anxiety, the kind that isn't bad enough for medication but is still troublesome.

My high school patient needed the full complement of treatments: medicine, exercise, and talk therapy. But for my patient Amy, physical activity alone helped tremendously, both in the moment and day to day, and it cleared the way for talk therapy to get at the underlying issues. Aerobic exercise complemented her yoga and provided the calm necessary to look inside and observe herself, rather than spend all of her emotional energy on the task of not being overwhelmed. She became much more aware of her own psychology and behavior than she had ever been. She came to recognize that there is a natural ebb and flow to her challenges and negative feelings, and to realize that she has to ride the waves — and that she can. Equally important, she noticed the improvement in herself and described it well: the divorce was like an earthquake that nearly reduced her life to rubble, but exercising had solidified her foundation; she knew there would be aftershocks, but she felt strong enough to withstand them.

It's astonishing how much Amy has changed. Her lawyer, her parents, her family therapist — even her husband, to some degree — have all commented that she seems like a different person. She's more in command of herself and her situation, refreshingly self-assured and realistically optimistic. The court battle may continue for years, but she's no longer overwhelmed by it, and exercise has been her best defense.

5
Depression
Move Your Mood

BILL DIDN'T KNOW he was missing out. When he turned fifty, he realized he was twenty pounds overweight, and he decided to go on a diet and start running. Before long, he began losing weight, and he noticed some dramatic side effects: he became less critical—of himself and of others—and less of a grouch. His wife and his children picked up on the shift and wanted to spend more time with him, which made him feel good and further improved his attitude. Bill had never had depression, per se, but there was no doubt he felt more passionate about life once he got into an exercise routine. He discovered, entirely by accident, that he could be happier.

Our understanding of depression followed a similar path. Pure serendipity led to our first antidepressants, when in the 1950s it was discovered that an experimental tuberculosis drug made people "inappropriately happy." A few years later, a new antihistamine produced similar mood-elevating effects and spawned a class of drugs called tricyclic antidepressants. Suddenly there were medical treatments that could soothe symptoms of depression. It was the first real hint at the radical notion that there might be a biological explanation for what had been considered an entirely psychological problem. This led to a search for how the brain controls the mind, and the entire landscape of the field changed.

In the fifty years since then, mood disorders have been the focal point of psychiatric research. We still don't know what causes depression, but we've made great strides in describing the brain activity underlying emotions. And the more we've learned about the biology of mood, the more we've come to understand how aerobic exercise alters it. In fact, it's largely through depression research that we know as much as we do about what exercise does for the brain. It counteracts depression at almost every level.

In Britain, doctors now use exercise as a first-line treatment for depression, but it's vastly underutilized in the United States, and that's a shame. According to the World Health Organization, depression is the leading cause of disability in the United States and Canada, ahead of coronary heart disease, any given cancer, and AIDS. About 17 percent of American adults experience depression at some point in their lives, to the tune of $26.1 billion in health care costs each year. It's impossible to know how many people try to commit suicide, but, tragically, in the United States someone succeeds about every seventeen minutes. For this reason, and also because 74 percent of depression patients experience some other disorder — including anxiety, substance abuse, and dementia — it's an urgent problem. Unfortunately, it's not getting any better.

One of the hurdles to conquering depression is that the disorder encompasses such a broad array of symptoms, most of which all of us experience at some point. Who doesn't feel grouchy, irritable, pessimistic, lethargic, apathetic, self-critical, or melancholy on occasion? Sadness, for instance, is a normal aspect of the human condition — a response to a loss. But being sad isn't the same as being depressed, unless the feeling never goes away or comes along with a certain number of the other symptoms.

And what's the difference between a symptom and a personality trait? My patient Bill had been critical and negative most of his life. He wasn't sick, technically, but he had what I would call a shadow syndrome of depression and was a perfect candidate for a lifestyle change to benefit his Eeyore-like outlook. Prescribing medicine to make people "better than well" is the subject of a long-standing ethical debate, and this is one area where exercise offers a huge advantage over antidepressants. Just because you don't have all the symptoms of depression doesn't mean you can't feel better. Bill's a happier guy because he started running. As I'll explain, the same would probably hold true if he were clinically depressed. Aerobic exercise has a positive impact on the entire range of depressive symptoms, regardless of whether they come individually, in the form of a mild episode, or conspire to form a disorder. Overall, I think of depression as an erosion of connections—in your life as well as between your brain cells. Exercise reestablishes those connections.

Within the spectrum of symptoms, there are distinctly different types of depression. I've had patients who don't eat and can't sleep, and others who eat too much and are so fatigued they feel like they can't get out of bed in the morning. Some can't make even the simplest decision, and they quietly withdraw from the world in a posture of helplessness, while others shout and challenge everything and everyone. Such contradictions make treatment difficult. If you have breast cancer, a biopsy can determine the best treatment. If you have depression, you may take a psychological test, and then it's a matter of trial and error to find a drug that might work; there is no blood test for depression.

Which brings us back to the quest to find a biological culprit. By reverse engineering our original, accidental antidepressants, we discovered that they increase the activity of the so-called monoamine neurotransmitters: norepinephrine, dopamine, and serotonin. When a research psychiatrist at Massachusetts Mental Health Center named Joseph Schildkraut found in 1965 that a breakdown product of norepinephrine, called methoxyhydroxyphenylglycol (MHPG), was reduced in depressed patients, we got excited about the idea that there was something to measure. If we could quantify the imbalance, we should be able to diagnose and attack the disease at the fundamental level of biology. His pioneering work led to the monoamine hypothesis, which holds that depression is caused by a deficit of these three neurotransmitters. Most of our treatments and research since then has been about trying to reverse that deficit.

THE NEW BUZZ

In 1970, fresh out of college, I landed a job at the Massachusetts Mental Health Center and walked right into this sea change happening in psychiatry. Schildkraut was a mentor to me, and I was lucky enough to have a firsthand look at the scientific inquiry into the biological theory of mood disorders. Two years later, I was off to medical school at the University of Pittsburgh, where I started my own daily psychoanalysis and steeped myself in the emerging brain science. Everyone at Pitt was already working on MHPG, so I settled for measuring lithium uptake into red blood cells as a possible way of identifying different mood problems. I also froze urine samples of patients with schizophrenia, which were then sent off to Linus Pauling at Stanford; learned how to program computers to conduct data analysis; and presented findings at a psychophysiology conference. Through my research, I was utterly caught up in the passion of trying to make psychiatry a "real" science.

Around the same time, I came across an article about a hospital in Norway that was offering depressed patients the option of treatment with antidepressants or daily exercise. This was a shock: These drugs were really just being introduced, and the results were forcing us to change the way we thought about treatment, yet here was a hospital offering exercise for people with severe depression. And it was working! But the results were lost in the shuffle. At a time when we were just plumbing the depths of the brain, the need was for hard science.

When I settled in Boston for my residency at the Massachusetts Mental Health Center, I landed in another epicenter—the running craze—just as it was taking off. We had Olympic gold medalist Frank Shorter competing against the world's best runners in his hometown marathon; we had Bill Rogers telling everyone to get out there; and we had a new phenomenon called the "endorphin rush."

A postdoctoral neuroscientist from Johns Hopkins University named Candace Pert had recently discovered that there were opiate receptors in the brain, meaning the body had a built-in way of killing pain with molecules that worked like morphine. Endorphins, as they became known, dulled pain in the body and produced euphoria in the mind. When elevated levels of endorphins were detected in the blood samples of a group of runners, everything seemed to fit. The theory that exercise fills your brain with this morphinelike substance matched the good feeling everyone got. It gave us the expression "runner's high," an extreme version of the effect. For me, it was the first time I connected the idea of exercise and mood.

Endorphins are considered stress hormones—and there are forty types of them, with receptors throughout the brain and body—that calm the brain and relieve muscle pain during strenuous exercise. They are the elixir of heroism, helping us ignore pain when we're physically overextended so we can finish the task at hand. Robert Pyles, the psychiatrist I mentioned in chapter 3, offers a good example. As a marathoner, he prided himself on always finishing, but that became a mighty challenge one year at Boston. His feet got tangled in a plastic trash bag someone had been using as a warm-up jacket near the starting line, and he hit the pavement, kneefirst. He popped back up and kept going, more shocked than anything. But as the miles ticked by, his stride started to feel funny, and by mile eighteen his swollen knee stopped working. He had fractured his femur. Every pounding step should have caused debilitating pain, but Pyles says he didn't notice it. It had to be the endorphins.

Pain is related to depression, and after Pert's discovery, others conducted experiments to see if endorphins were indeed the link between exercise and elevated mood. They expected to find that endorphin-blocking drugs would prevent runner's high, but there were conflicting results. Then we found that endorphins produced in the body—the ones detected in the runners—cannot pass into the brain, and scientific enthusiasm for the endorphin rush faded. Endorphins obviously weren't the single answer, so they were abandoned in the lab. Now we're coming back around to them. Studies suggest that endorphins produced directly in the brain contribute to the general feeling of well-being that usually comes along with exercise. The truth is, how much they contribute to the magic is uncertain.

The problem with the strictly biological interpretation of psychology is that we sometimes lose sight of the fact that the mind, brain, and body all influence one another. In addition to feeling good when you exercise, you feel good *about* yourself, and that has a positive effect that can't be traced to a particular chemical or area in the brain. If you've been feeling down and you start to exercise and feel better, the sense that you're going to be OK and that you can count on yourself shifts your entire attitude. The stability of the routine alone can dramatically improve your mood. Clearly, there's something going on.

One of the best examples is a landmark research project from the Human Population Laboratory in Berkeley called the Alameda County Study. Researchers tracked 8,023 people for twenty-six years, surveying them about a number of factors related to lifestyle habits and healthiness starting in 1965. They checked back in with the participants in 1974 and in 1983. Of all the people with no signs of depression at the beginning, those who became inactive over the next nine years were 1.5 times more likely to have depression by 1983 than their active counterparts. On the other hand, those who were inactive to begin with but increased their level of activity by the first interval were no more likely to be depressed by 1983 than those who were active to begin with. In other words, changing your exercise habits changes your risk for depression.

Several other sweeping studies have looked at the correlation from slightly different angles, and all of them came to the same conclusion. A massive Dutch study of 19,288 twins and their families published in 2006 showed that exercisers are less anxious, less depressed, less neurotic, and also more socially outgoing. A Finnish study of 3,403 people in 1999 showed that those who exercise at least two to three times a week experience significantly less depression, anger, stress, and "cynical distrust" than those who exercise less or not at all. This was a cardiovascular risk factor survey that included questions about mood, which is to say they were talking about a broader range of symptoms than clinical depression. Another study, from the epidemiology department at Columbia University published in 2003, surveyed 8,098 people and found the same inverse relationship between exercise and depression.

CONVERGING PATHS

When the blockbuster drug Prozac came along, it was the first antidepressant that corrected the chemical imbalance of just one of the suspect neurotransmitters. Prozac is the mother of a class known as selective serotonin reuptake inhibitors (SSRIs), which prevent serotonin from being recycled at the synapse, leaving more available for use, and theoretically restoring normal transmission in the brain. Prozac was exciting because it worked for a lot of people and it pointed to a single problem that could be fixed. It had a huge impact because it didn't just squelch negativity, it boosted self-esteem, which is a different dimension of the disease.

Twenty years later, it's evident that Prozac and its SSRI daughters don't work for everyone, nor do antidepressants that target norepinephrine, dopamine, or any combination of the three. One of the issues is side effects. To take an example, a good portion of my patients on SSRIs develop problems with their sex lives after a few months. Some estimates suggest that more than 50 percent complain of sexual side effects, ranging from lack of interest to lack of functionality. (It says something that SSRIs are often used as a treatment for premature ejaculation and for sexual offenders.) Sexual problems are easy for people to miss or dismiss, especially when they're otherwise feeling good, but they can sneak up on us and lead to other issues. Sexual feelings and passions are primary drivers in all of us, and muzzling them can leave us with a general lack of passion for life or a lack of intimacy or maybe a list of missed opportunities. On balance, the side effects are certainly outweighed by the most serious consequences of depression, but they get in the way for a lot of people. Now SSRIs carry a warning that they may increase the risk of suicidal thoughts and actions in children and adolescents, a finding that is still in question. And stories are coming out about the difficulty of withdrawing from this class of drugs, especially from venlafaxine (Effexor).

Recently, I started treating a successful entrepreneur whose life was a mess. He and his wife were separated because he'd had an affair, and he'd also lost his business. He came to me to find out more about an issue that had cropped up in couple's therapy, where it became clear that he suffered from ADHD.

Dead set against the idea of putting anything "unnatural" into his body, he resisted medication. But he finally agreed to try a stimulant, mostly because his wife was pressuring him, and he felt terribly guilty about cheating. We tried several different drugs, but we quickly stopped all of them. He got headaches, stomachaches, muscle pain.

His attention problems aside, the primary issue, I told him, was that he was depressed. He was inactive, unmotivated, and he felt hopeless. He had done nothing to sort out his work situation, and although it had been months, he denied there was a problem. Then one day, he came in, and I could see his demeanor had deteriorated. Typically very neat, he was unshaven and unkempt and told me he was having a hard time getting out of bed.

I insisted that he try an antidepressant, and I put him on the SSRI escitalopram (Lexapro). He had a severe reaction—he felt nauseated and started throwing up—and said he didn't want to try another.

He'd been physically active in the past, and I told him that he should really be exercising every day. I'd been talking about it all along, but after the Lexapro episode, I explained what an enormous effect exercise has on the brain, and I gave him several relevant studies in an appeal to his professional sensibility.

Two weeks later, he looked like a different man. He was smiling and confident, and he felt good about the fact that he had been running nearly every day. Over the course of the next month, I saw him take up his job search in earnest and make strides in reconciling with his wife. And for the first time, he said he was hopeful that they might get back together. More than anything, he was amazed that he could feel so different and sustain that feeling.

Aside from elevating endorphins, exercise regulates all of the neurotransmitters targeted by antidepressants. For starters, exercise immediately elevates levels of Schildkraut's favorite neurotransmitter, norepinephrine, in certain areas of the brain. It wakes up the brain and gets it going and improves self-esteem, which is one component of depression.

Exercise also boosts dopamine, which improves mood and feelings of wellness and jump-starts the attention system. Dopamine is all about motivation and attention. Studies have shown that chronic exercise increases dopamine storage in the brain and also triggers the production of enzymes that create dopamine receptors in the reward center of the brain, and this provides a feeling of satisfaction when we have accomplished something. If the demand is there, the dopamine genes get activated to produce more, and the overall effect is a more stable regulation of these pathways, which are important to controlling addictions.

Serotonin is equally affected by exercise, and it's important for mood, impulse control, and self-esteem. It also helps stave off stress by counteracting cortisol, and it primes the cellular connections in the cortex and hippocampus that are important for learning.

THE TRUE TEST

We've known for a while that exercise influences the same chemicals as antidepressants do, but nobody had done a scientifically sound head-to-head comparison until researchers at Duke University took up the task in 1999. In a landmark study affectionately called SMILE (Standard Medical Intervention and Long-term Exercise), James Blumenthal and his colleagues pitted exercise against the SSRI sertraline (Zoloft) in a sixteen-week trial. They randomly divided 156 patients into three groups: Zoloft, exercise, or a combination of the two. The exercise group was assigned to supervised walking or jogging, at 70 to 85 percent of their aerobic capacity, for thirty minutes (not including a ten-minute warm-up and a five-minute cool-down) three times a week. The results? All three groups showed a significant drop in depression, and about half of each group was completely out of the woods—in remission. Another 13 percent experienced fewer symptoms but didn't fully recover.

Blumenthal concluded that exercise was as effective as medication. This is the study I photocopy for patients who are skeptical of the idea that exercise changes their brain chemistry enough to help their depression, because it puts the issue in terms that are as black-and-white as psychiatry can hope to deliver, at least for now. The results should be taught in medical school and driven home with health insurance companies and posted on the bulletin boards of every nursing home in the country, where nearly a fifth of the residents have depression. If everyone knew that exercise worked as well as Zoloft, I think we could put a real dent in the disease.

Reading between the lines of the SMILE study gets at the complex issues that have kept exercise from being accepted as a medical treatment. Just as with Andreas Broocks's comparison of exercise to the antianxiety drug clomipramine in 1997, although both groups ended up at the same point of improvement, the patients taking medication experienced more immediate relief. At first glance, this seemed to contradict the cautions from pharmaceutical companies that antidepressants might take three weeks to work. But those time estimates are based on statistics; I have had plenty of patients over the years who have responded within a few days.

Conversely, what about the studies showing that a single bout of exercise can improve mood? For instance, in 2001, Northern Arizona University psychology professor Cheryl Hansen showed that as little as ten minutes of exercise can immediately improve vigor and mood in healthy subjects. But if Hansen had reexamined mood several hours later, she probably would have found that it had returned to baseline. So while it's important for us to realize that we can lift our mood with one session of exercise, it's also important to keep in mind that it takes longer to change our mood day to day.

In Blumenthal's study, he assessed mood once a week, before exercise. He did see immediate improvement in some patients, but not as dramatically as with medicine. A crucial aspect of recovering from depression is being able to predict that you're still going to feel good five minutes from now, and then five hours from now. Eventually you'll be confident that you'll feel better tomorrow morning. That might take a little longer with periodic exercise.

Six months after the study, Blumenthal and his colleagues surveyed the patients to see how they were doing and found that exercise worked even better than medicine over the long term. About

30 percent of the exercise group remained depressed versus 52 percent of those on medication and 55 percent for those in the combined treatment group. (Blumenthal has some interesting theories about what happened in the combo group, which I'll explain in a bit.) And of the patients who were in remission after the initial study, just 8 percent of the exercise group had a relapse versus 38 percent in the medication group—quite a significant difference.

After the initial four-month trial, the subjects were free to seek treatment (or not) however they chose, which complicated the results. Some signed up for psychotherapy, some in the medication group began exercising, and some in the exercise group started taking medicine, which brings a lot of variables into play. But Blumenthal's group found that the most significant predictor of whether someone felt better was how much they exercised. Specifically, every fifty minutes of weekly exercise correlated to a 50 percent drop in the odds of being depressed. Blumenthal stopped short of concluding that exercise *caused* remission. Maybe the reverse was true: patients who continued to exercise did so *because* they were less depressed. It's the classic chicken-egg problem scientists face when correlating physical activity and mood. Does it really matter whether you're less depressed because you exercise or whether you exercise because you're less depressed? Either way, you feel better.

But how to account for the combination group's unexpected results? Blumenthal assumed those who exercised and took Zoloft would show the best results, but they had the worst relapse rates. He speculated that some of the participants didn't like the idea of antidepressants, especially considering they had expressly signed up for a study to test the effects of exercise on depression. Several were disappointed when they found out they would also receive medication, and during the trial some said they thought the medication was *interfering* with the benefits of exercise. That's unlikely from a physiological standpoint, but it's possible that taking medication undercut the sense of self-mastery that exercise confers. As

Blumenthal explained in the study, "Instead of incorporating the belief 'I was dedicated and worked hard with the exercise program; it wasn't easy, but I beat this depression,' patients might incorporate the belief 'I took an antidepressant and got better.'"

THE BEST FIX

When we talk about depression, we don't use the word *cure* because we only have subjective measures for behavior and emotions. About a third of patients with depression achieve full remission of their symptoms with antidepressants. Another third feel much better with medication, but they may continue to have problems with motivation, lethargy, and fatigue. The bad thoughts are gone, but even though they're able to get out of bed, they're still not knocking on the door for a new job or doing what they should be doing. They're feeling less than well, lingering in the shadows of depression. The current *Diagnostic and Statistical Manual of Mental Disorders* lists nine symptoms for depression, and you need to have six to receive a diagnosis of depression. Say you can't concentrate, can't sleep, feel worthless, and aren't interested in anything. That's four. Technically, you aren't depressed. What are you, then? Just miserable? My point is, with any degree of depression, you need to snuff it out completely. And exercise is starting to be taken very seriously in this regard.

Madhukar Trivedi, a clinical psychiatrist who is the director of the Mood Disorders Research Program at the University of Texas Southwestern Medical School, has been researching the effectiveness of using exercise to augment antidepressants. In 2006 he published a pilot study showing that patients who weren't responding to antidepressants lowered their scores on a common depression test by 10.4 points on a 17-point scale — a huge drop — after twelve weeks of exercise. All seventeen patients were deeply depressed and had been taking antidepressants for at least four months. The drugs weren't working, but they stayed on them for the experiment.

Trivedi enlisted the Cooper Institute to set up the exercise protocol. Then they let the patients, who were inactive to begin with, exercise at home, either walking or riding a stationary bike as often and as rigorously as they wanted. The only requirement was that they expend a certain amount of energy each week. Most chose to walk an average of fifty-five minutes three days a week. Nine people dropped out, which isn't unusual, but of the eight who completed the exercise program, five achieved full remission. Even the patients who stuck with the program for only a few weeks showed an improvement in symptoms.

The numbers were small, but Trivedi's results were huge. At least for some people, when drugs don't work, exercise does. It begs the question, why wouldn't you add exercise right from the beginning? Especially if you're playing trial and error with a host of other medications? But the allure of the magic pill is powerful, and it takes a long time to overturn attitudes. Just ask T. Byram Karasu, who was in charge of the American Psychiatric Association's work group on major depressive disorders. He pushed to get the APA to formally adopt exercise in its treatment guidelines for depression and suggested that psychiatrists tell every patient to walk three to five miles a day or do some other type of vigorous exercise. The APA balked, presumably because, while most doctors acknowledge the anecdotal evidence that exercise improves mood, they say there isn't enough scientific evidence. In this age of deconstructing the brain and unlocking the secrets of the life and death of cells, it's difficult for psychiatrists to consider such a holistic strategy as exercise a treatment.

Any doctor will tell you that the worst patients are other doctors. Imagine the difficulty, then, in convincing a patient of mine with a medical degree to exercise for her depression. Grace, who has a history of mild bouts of depression, also happens to be a psychiatrist with a sophisticated knowledge of medicine. Even so, we haven't been able to find an antidepressant without side effects. SSRIs seem to work best for her, but she would try one and then stop because they all made her gain weight very quickly. She's brilliant and well aware of the biology of exercise—at least some of it—but she just didn't do it.

Last summer, she injured her back and ended up immobilized in bed for a period. Purely for physical rehabilitation, she started swimming. It was the only thing she could do, and it felt good because the water supported her body and eased her pain. She enjoyed it and began exercising in the pool for three hours a day. Not only did her pain decrease, but she also started seeing some long-lost muscle tone and she felt good about herself.

Then the pool shut down for the winter, her back flared up, and her mood took a dive again. This time she also got angry. Flat on her back with few options for exercise, she started lifting weights—just three-pound dumbbells that she would pump up and down fast enough to get her heart rate up several times a day. Even that minor activity helped, and more important, the experience sparked a shift in her brain and in her mind. I've been seeing Grace for years, but there was something about this situation that put exercise into perspective for her.

She's worked out the kinks in her back and has been very steady with her swimming routine. She says she's able to think and write more creatively, and she feels a newfound sense of vigor that hasn't gone unnoticed by her family and friends. Not that it should have surprised her: She recalled that when she was training every day for the tae kwon do team in college, she did some of her best work. Then as a young doctor in Boston, she got into running marathons. Like many people, it was only after she started a family that she fell out of the habit of exercise. "I just got too busy and forgot the benefits of working out," she says. "Now I feel like I have my brain back."

HOLES IN THE THEORY

It wasn't until we were able to take good pictures of the brain that we really began to understand how various medical treatments and exercise break us out of the bonds of depression. In the early 1990s, using MRI, we noticed bright patches on brain scans of certain depressed patients. The hyperintensities, as they are known, appeared in the white matter, the portion of densely bundled axons that connect neurons in the gray matter of the cortex. Zooming in further, we found differences in the volume of the cortex — the gray matter was physically shrunken. Gray matter is the thin, wrinkly covering of the brain made up of cells that direct all of our complex functions such as attention, emotions, memory, and consciousness. The MRI scans pointed toward a radical notion: that chronic depression may cause structural damage in the thinking brain.

Related research showed that depressed patients also had measurable changes in the amygdala and the hippocampus, crucial players in the stress response. We knew the amygdala was central to our emotional life, but we were just discovering that the memory center was also involved in stress and depression. In 1996 Yvette Sheline of Washington University in St. Louis compared ten patients with depression to ten healthy controls of the same physical stature and educational background, and she found that the hippocampus of depressed patients was up to 15 percent smaller compared to that of the controls. She also found evidence that the degree of shrinkage was directly related to the length of depression, and this was news. It might explain why so many patients with depression complain of learning and memory trouble, and why mood deteriorates in Alzheimer's, the neurodegenerative disease that begins with erosion of the hippocampus.

High levels of the stress hormone cortisol kill neurons in the hippocampus. If you put a neuron in a petri dish and flood it with cortisol, its vital connections to other cells retract. Fewer synapses develop and the dendrites wither. This causes a communication breakdown, which, in the hippocampus of a depressed brain, could partly explain why it gets locked into thinking negative thoughts—it's recycling a negative memory, perhaps because it can't branch out to form alternative connections.

Neuroimaging has cracked open an entirely new view of the biology of depression. The picture was—and still is—crude and fuzzy, but now positron-emission tomography (PET) scans and functional MRI (fMRI) scans allow scientists to go beyond snapshots and see the brain at work. At the same time, we learned that new nerve cells are born every day in the hippocampus and possibly in the prefrontal cortex—two areas shriveled in depression. The new tools and new discoveries led to a reformulation of the neurotransmitter theory.

Our understanding hasn't junked the old theory, just expanded it. Now we see depression as a physical alteration of the brain's emotional circuitry. Norepinephrine, dopamine, and serotonin are essential messengers that ferry information across the synapses, but without enough good connections in place, these neurotransmitters can only do so much. As far as the brain is concerned, it's job is to transfer information and constantly rewire itself to help us adapt and survive. In depression, it seems that in certain areas, the brain's ability to adapt grinds to a halt. The shutdown in depression is a shutdown of learning at the cellular level. Not only is the brain locked into a negative loop of self-hate, but it also loses the flexibility to work its way out of the hole.

Redefining depression as a connectivity issue helps explain the wide range of symptoms people experience. It's not just a matter of feeling empty, helpless, and hopeless. It affects learning, attention, energy, and motivation—disparate systems that involve different parts of the thinking brain. Depression also affects the body, shutting down the drive to sleep, eat, have sex, and generally look after ourselves on a primitive level. Psychiatrist Alexander Niculescu sees depression as a survival instinct to conserve resources in an environment void of hope — "to keep still and stay out of harm's way," he wrote in a 2005 article in *Genome Biology*. It's a form of hibernation: When the emotional landscape turns wintry, our neurobiology tells us to stay inside. Except that it can last much longer than a season. It's as if our entire being has said, there's nothing out there for me, so I may as well quit. Thus the atrophy, the halt of neuroplasticity, neurogenesis, and overall lack of connectedness. It's no wonder we haven't been able to define depression as a single problem.

THE UNDERLYING CONNECTION

If depression is primarily a communication breakdown, or a loss of the brain's ability to adapt, it's very good news for the value of exercise. We learned in the early 1990s that brain-derived neurotrophic factor (BDNF) protects neurons against cortisol in areas that control mood, including the hippocampus. It is the fertilizer that encourages neurons to connect to one another and grow, making it vital for neuroplasticity and neurogenesis. While extremely high levels of cortisol can decrease BDNF, antidepressants and especially exercise do just the opposite. BDNF is the rope in a tug of war between chronic stress and adaptability. The Miracle-Gro molecule became the new serotonin, and we began measuring, blocking, increasing, and otherwise tweaking it in every way imaginable, to see how it affected mood in mice and men.

We can't ask a rodent if it's depressed, but we can see how it reacts to inescapable stress. If its feet are shocked, does it attempt to escape or does it freeze? This is the experimental model for learned helplessness, a popular way of describing human depression that implies an inability to cope with adversity and to take the action necessary to survive and flourish. If the lab animal gives up, it's considered to be depressed.

In one such experiment, when BDNF was injected directly into the hippocampi of mice, they were much quicker to try to escape than their nontreated counterparts. The injection seemed to have the same effect as exercise and antidepressants on the animals' behavior. Conversely, scientists have bred mice with genes that produce 50 percent less BDNF and found that they don't respond well to antidepressants, suggesting it's a necessary ingredient for the drugs to work. Such mice were significantly slower to try to escape stress than their peers with normal BDNF function.

In people, scientists are limited to measuring BDNF in the bloodstream, which, at best, offers a rough approximation of levels in the brain. One study of thirty depressed patients showed that every one of them had lower than normal BDNF levels. In another, antidepressants restored depressed patients' BDNF levels to normal, and yet another showed that higher levels correspond to fewer symptoms. In postmortem studies of people with depression who died of suicide, their brains had significantly decreased levels of BDNF. Even in healthy people, low BDNF levels have been correlated with personality traits that make them more vulnerable to depression, such as neuroticsm and hostility.

Exercise boosts BDNF at least as much as antidepressants, and sometimes more, in the rat hippocampus. One study showed that combining exercise with antidepressants spiked BDNF by 250 percent. And in humans, we know that exercise raises BDNF, at least in the bloodstream, much like antidepressants do.

As with norepinephrine in the 1960s, BDNF may be the tip of the iceberg. Today, research focuses on BDNF as well as vascular endothelial growth factor (VEGF), fibroblast growth factor (FGF-2), insulin-like growth factor (IGF-1), and all the related chemicals involved in encouraging neuroplasticity and neurogenesis. At the same time, pharmaceutical companies are funding research to mark and measure all these factors, and to map the genes affected by them, so they can figure out how to mimic their actions. BDNF and its neurotrophic brethren are much farther upstream in the neurochemical cascade than serotonin, closer to the source. Ultimately, the genes have to turn on the flow.

The shift from the neurotransmitter hypothesis to the connectivity theory is a move from outside to inside the nerve cell. In addition to working at the synapse, as serotonin does, BDNF turns on genes to produce more neurotransmitters and neurotrophins, puts the brakes on self-destructive cellular activity, releases antioxidants, and provides the proteins used as building material for axons and dendrites. These gene-regulated adaptations of BDNF might go a long way toward explaining the delayed effect of antidepressants.

It can often take three weeks for antidepressants to work. Is it merely a coincidence that the process of neurogenesis—from the time a stem cell is born in the hippocampus until it plugs into the network—takes approximately the same amount of time? Many researchers think not. The latest twist to the connectivity theory is that a shutdown of neurogenesis might be a factor in depression. Some have shown that blocking neurogenesis in rats negates the effect of antidepressants, so it's a possibility. This could provide an even stronger link between exercise and its antidepressant effect, given that exercise clearly elevates BDNF and its sibling growth factors, and that these are essential to the building process of neurogenesis. Nobody has shown that a deficit of BDNF *causes* depression, but it's not for lack of trying. In 1997 Yale psychiatrist Ronald Duman published "A Molecular and Cellular Theory of Depression" in the *Archives of General Psychiatry,* and ever since he and others have been unraveling the story of BDNF. In 2006 he charted how various treatments influence BDNF, including all available antidepressants, as well as less common forms of treatment such as electroconvulsive therapy (ECT) and transcranial magnetic stimulation (TMS). They all boost BDNF in the hippocampus, and Duman pointed out that ECT, the most effective treatment, raises it by 250 percent.

But how is it that a blunt physical manipulation such as ECT—which sends an electrical current through the brain to induce a seizure—can work like drugs, therapy, and exercise? I think ECT provides a useful metaphor. If we look at depression as a sort of brain lock, then we can see a common thread between these approaches: They are all shocks of a sort. They send sparks flying to change the dynamic in the brain. Some parts are stuck in a constant whir, and others are locked in place. The key, I think, is to wake up the brain and the body so you can pull yourself out of the downward spiral. What makes aerobic exercise so powerful is that it's our evolutionary method of generating that spark. It lights a fire on every level of your brain, from stoking up the neurons' metabolic furnaces to forging the very structures that transmit information from one synapse to the next.

BREAKING THE BONDS

While molecular scientists approach brain lock with tools to pick it, Emory University neurologist Helen Mayberg wants to smash it wide open. Several years ago, she tested a radical therapy called deep brain stimulation (DBS), in which she inserted an electrode into the subgenual cortex in a half dozen severely depressed patients, for whom every other form of medical treatment had failed. "These people are stuck," Mayberg says. "They have an inability to put thought to action because they're not in gear. We need to find a way, metaphorically, to unstick them." She used voltage, and the results were profound: All six patients spontaneously described sensations such as a "disappearance of the void" right there on the operating table, the second the electrodes were switched on. Four of them eventually achieved full remission.

The area of the brain Mayberg targeted is the tip of the anterior cingulate, which is the major way station between information coming down from prefrontal cortex and up from limbic system—the landing between upstairs and downstairs in your emotional stairwell. As the seat of executive function, it prioritizes what you pay attention to and indirectly regulates the limbic system, integrating cognitive and emotional signals. If it fails to shift attention from the negative, in the case of depression, you can't think about anything else. "Not being able to initiate, not being able to think clearly, not being able to care about your family — those are secondary to something that your attention is directed to internally, but that is a false signal," Mayberg explains. "You can fix it, and then other problems may be more accessible."

The real goal of DBS is to bring the prefrontal cortex back online so you can use your executive function to tackle the core issues. It frees you up to attack the problem rationally — *I'm not a bad person; my children love me; and my life is not beyond repair*. This is also one of the effects of exercise. In 2003 a group of German neuroscientists did an experiment with twenty-four people with mild depression who were being treated with drugs, and ten healthy controls. They all took a battery of neuropsychological tests designed to gauge executive function. They repeated the tests after thirty minutes of stationary cycling at 40 percent of their maximum heart rate, then again at 60 percent — low enough to keep lactic acid out of the equation. The depressed patients showed significant improvement on two of the four tests at both intensity levels, which demonstrated that exercise immediately improves the highest form of thinking. Even one bout of exercise was enough to influence the prefrontal cortex. The ten nondepressed volunteers didn't show much improvement, but then, there was nothing to fix.

Mayberg would be the first to say that executive function is still only part of the story. In comparing PET scans of patients who responded to antidepressants against others who responded to cognitive behavior therapy, she found that the two approaches change activity levels of the limbic system from opposite directions. Antidepressants seem to work through a bottom-up chain of events, meaning the activity begins in the brain stem and ripples through the limbic system until it reaches the prefrontal cortex. This might explain why antidepressants relieve the physical effects first—we feel more energetic before we feel less sad. With cognitive behavioral therapy and psychotherapy, we feel better about ourselves before we feel better physically. Therapy works from the prefrontal cortex down, to modify our thinking so we can challenge the learned helplessness and spring ourselves out of the hopeless spiral.

The beauty of exercise is that it attacks the problem from both directions at the same time. It gets us moving, naturally, which stimulates the brain stem and gives us more energy, passion, interest, and motivation. We feel more vigorous. From above, in the prefrontal cortex, exercise shifts our self-concept by adjusting all the chemicals I've mentioned, including serotonin, dopamine, norepinephrine, BDNF, VEGF, and so on. And unlike many antidepressants, exercise doesn't selectively influence anything—it adjusts the chemistry of the entire brain to restore normal signaling. It frees up the prefrontal cortex so we can remember the good things and break out of the pessimistic patterns of depression. It also serves as proof that we can take the initiative to change something. This paradigm holds true for exercise's effect on mood in general, regardless of whether we're depressed or coping with some nagging symptoms. Or even if we're just having a bad day.

The idea of attacking depression from both directions couldn't be more important. "Even with the brain stimulation patients, once you get the system back online, they need mental rehabilitation," Mayberg says. "Part of the initial rehab is simply to get them doing something. The best kind of behavioral therapy is to just go outside, go for a walk. Do something. It doesn't require you to have elaborate planning. It doesn't put you into a negative spin. If you do a physical activity, it's very quickly reinforcing because, before, you couldn't initiate to make yourself do something."

If your prefrontal cortex has been offline for a while, you need to reprogram it, and exercise is the perfect tool. You begin to look at the world differently, and you see trees instead of a barren wasteland. When you see yourself moving, that alone is an achievement — proof that you can help yourself.

OUT OF THE TUNNEL

Science has come a long way since our search for the single culprit began, and the decades' worth of research generated from the monoamine hypothesis has taught us volumes about the biology of emotions. The closer we get to the cause of depression, the more complex it appears. When we began, everyone was fairly certain that the problem was an imbalance of neurotransmitters at the synapses. Now we know for certain that it's not so simple.

Ironically, I think this is precisely why exercise has yet to be embraced as a medical treatment. It doesn't simply raise serotonin or dopamine or norepinephrine. It adjusts all of them, to levels that, we can only presume, have been optimally programmed by evolution. The same goes for exercise's effect on BDNF, IGF-1, VEGF, and FGF-2, which provide the building material and oversight for the construction of new connections and neurons. In short, exercise affects so many variables in the brain that its nigh impossible to isolate its effect as we'd like — in the name of hard science. But the evidence is there, from the action of microscopic molecules to massive surveys of tens of thousands of people over the years. Yes, exercise is an antidepressant. But it is also much more.

Still, it's not surprising that about half the patients in studies on exercise and depression drop out. Perhaps because most of them are inactive to begin with, getting started is that much more challenging. This is critical for doctors to keep in mind when recommending exercise. With people who are already feeling hopeless, it's important to keep expectations reasonable, so as not to reinforce the negative. On the other hand, studies have shown that even people who find the act of exercise inherently unpleasant have a positive mood swing the minute the workout is over. If you know what's on the other side, it's easier to push through the difficulty.

Human beings are social animals, so if you're depressed, it would be ideal to choose a form of exercise that encourages making connections and that can take place outside or in some environment that stimulates the senses. Asking someone to join you in exercise and putting yourself in a new setting will give those newly hatched neurons a powerful reason for being; new connections need to be formed to represent the sensory stimulation. Breaking out of the emptiness that the brain has been locked into provides a sense of purpose and self-worth that evokes a positive future. Once you develop the positive feeling, you need to devote it to something. Then you get the bottom-up motivation and physical boost combined with the top-down reevaluation of yourself. By motivating the body to move, you're encouraging the mind to embrace life.

THE PRESCRIPTION

The first question people ask me when I suggest exercise as therapy is "How much should I do?" There is no firm answer — especially given depression's wide spectrum of symptoms and severity — but Madhukar Trivedi has drawn some conclusions about the amount necessary to be effective. By quantifying exercise as a dose, he hopes to cast treatment in terms the medical profession might accept. This is all-important, because doctors have to spend time with their patients to figure out what sort of regimen is likely to keep them moving.

In one study, Trivedi and Andrea Dunn divided eighty depressed patients into five groups, four with exercise protocols of different intensities and frequencies and one control group assigned to supervised stretching only (to see if social interaction with the supervisors had any effect). They used calories burned per pound of body weight as the "dose" measure. The high-intensity groups burned an average of 1,400 calories (eight calories per pound) during the course of either three or five sessions per week. At the end of three months, regardless of frequency, the high-intensity groups cut their depression scores in half. Practically speaking, their symptoms dropped significantly. The low-intensity groups burned an average of 560 calories (three calories per pound) and lowered their scores by a third, about the same as the stretching group—and about as effective as a placebo. What this means for us is what I always tell people: some exercise is good, more is better (to a point).

Trivedi and Dunn based the high dose on public health recommendations for exercise, which suggest thirty minutes of moderate aerobic activity on most days. If you weigh 150 pounds, that would translate into about three hours at moderate intensity per week. The low dose would be equivalent to eighty minutes per week.

Just multiply your body weight by eight to figure out how much you should be burning for the high dose, and then head to the gym to find out how many calories you burn during a given workout (most of the aerobic machines track this for you). If you weigh 150 pounds and burn 200 calories in thirty minutes on the elliptical trainer, you'd want to do six sessions a week to meet the high dose.

I would recommend at least this much exercise for anyone who, like my patient Bill, has a shadow syndrome of depression. These are people who aren't clinically depressed but who tend to look at life with a primarily pessimistic attitude, or who have the impression that nobody in the world, themselves included, meets their high standards. For Bill, when he got into running and lifting weights, he met up with a group of regulars at the gym every morning for a workout and coffee. His performance and relationships at work improved, and he felt like he didn't look at every new project with knee-jerk resentment. He almost welcomed new challenges, which completely changed his wife's view of him.

Then there are those people who are just moody. They tend to have low self-esteem, partly because they're so uncertain about how they'll feel from day to day — crabby or buoyant? One of my patients, Jillian, fit this description. We started exploring this issue after she got engaged to the man of her dreams and began feeling blue and irritable more often than not. Naturally, I had been pushing her to exercise, and she finally joined a gym near her office. Wisely, she teamed up with a co-worker who was already going, and they supported each other to work out every day at lunchtime. After a few months, she was pleased with herself, and she expressed how stabilizing the daily activity was. It helped her maintain a rhythm in her life, which added to her sense of stability.

Some people who have mood swings may describe themselves as manic-depressive or bipolar, but that's a problem of an entirely different magnitude. I haven't discussed bipolar disorder primarily because so little research has been done on exercise's effects on it. Recently, though, a preliminary study clearly showed that hospitalized patients with bipolar disorder who participated in a walking program suffered fewer depressive symptoms and less anxiety compared to those who were unwilling or unable to participate. And giving bipolar patients a stable social routine has been shown to improve long-term outcomes. It's only been very recently that exercise started making its way into treatment protocols for bipolar disorder.

In some ways, exercise is even more important for prevention than it is for treatment. One of the first symptoms of depression, even before your mood drops to new lows, is sleep disturbance. Either you can't get up or you can't get to sleep or both. I think of it as sleep inertia — trouble starting or stopping. First you lose your energy, then your interest in things. The key is to get moving immediately. And do not stop. Set up a schedule for a daily walk, run, jog, bike ride, or dance class. If you can't sleep, go for a walk in the dawn light and do it every day. Take the dog out, change your schedule—run from the depression. Burn those 1,400 calories as if your life depended on it, and nip it in the bud.

If you have severe depression, you may feel like you're at the bottom of a pit, in a state of slow death, and that it's almost impossible to make it outside or go to the gym or even think about moving. First, see your doctor about medication and get yourself some omega-3 supplements, which are proven to have antidepressant effects. This will, I hope, loosen up the brain lock enough for you to at least go for a walk. Ask for help. Get a friend or family member to agree to come by every day, at the same time if possible, to escort you outside and around the block. In England and Australia, walking groups for patients with depression have been popular for years, and now they're starting to make their way to the United States, so check the Internet to see if there is one in your area. If that's not an option, and you have the means, set up a regular time with a personal trainer. This might sound crazy, I realize, if you feel like you can't even lift yourself off the couch, but if that's the case, activity is all the more urgent.

Exercise is not an instant cure, but you need to get your brain working again, and if you move your body your brain won't have any choice. It's a process, and the best strategy is to take it one step—and then one stride—at a time. Start slowly and build on it. At its core, depression is defined by an absence of moving toward anything, and exercise is the way to divert those negative signals and trick the brain into coming out of hibernation.

6
Attention Deficit
Running from Distraction

"I SUSPECT THE first time I realized I was not like my peers was at the early age of three when I discovered no one else in my family or neighborhood was forced to wear a child leash," wrote Sam, a thirty-six-year-old venture capitalist who came to me in hopes of understanding his lifelong disorder, which was starting to manifest itself in his young son. "I have always been known in the family as the troublemaker and spent most of my childhood in the doghouse and the 'dunce' corner. My teachers felt that I had the ability to be a good student but never fully applied myself. I am able to express myself well and organize my thoughts, but often procrastinate."

Sam is no dunce, but like so many others with attention-deficit/hyperactivity disorder (ADHD), his erratic behavior led everyone around him to label him as stupid or stubborn or spoiled. He didn't want his son to suffer the same shame, and he was now seeking help at the encouragement of his business partner and his wife. "Neither quite understands how I can function with so much chaos in my life," he told me.

Chaos, high drama, deadline pressure—acute stress of any form acts like a drug for Sam's brain. His letter to me outlining his history acknowledges that he had disciplinary problems because he didn't deal well with authority figures and that he got into drugs at age fourteen. Yet he wasn't exactly a delinquent. When he turned sixteen, his parents forbade him from getting his driver's license until he shaped up, and he boosted his GPA from 1.5 to 3.5 almost overnight. Proof, many would argue, that his teachers were right: he just needed to try.

But the problem with Sam wasn't his attitude. ADHD stems from a malfunction of the brain's attention system, a diffuse linkage of neurons that hitches together areas controlling arousal, motivation, reward, executive function, and movement. Let's take one element of the attention system: motivation. While it's true that people with ADHD "just need to get motivated," it's also true that, like every other aspect of our psychology, motivation is biological. What about the child who can't pay attention in class but can sit perfectly still for hours playing a video game? Or the woman who "spaces out" when her husband is talking but has no trouble focusing on magazine gossip about Brad and Angelina? Obviously, they can pay attention when they want to, right? Not exactly. If we were to look at functional MRI (fMRI) scans of the brains of these people — and scientists have — we would see distinct differences in activity at the reward center in each situation. The reward center is a cluster of dopamine neurons called the nucleus accumbens, which is responsible for doling out pleasure or satisfaction signals to the prefrontal cortex, and thus providing the necessary drive or motivation to focus.

The sort of stimulation that will activate the reward center enough to capture the brain's attention varies from person to person. What wound up working for Sam was the rigid structure and rigorous physical activity of college athletics — and the desire to prove to everyone back home that he wasn't stupid. He put himself through school by playing Division III football and lacrosse and made the dean's list a number of times. "I believe participating in a sporting regimen which required five a.m. practice sessions," he wrote, "was the turning point that allowed me to see that I could function better in all my endeavors."

Now he runs several miles every morning and is a partner in a venture capital firm, matching entrepreneurs with big-money investors. In the parlance of this rarefied realm, Sam is what's known as a rainmaker—a high-energy personality with the social skills and business savvy to make deals happen. When there's a big one on the table, he has no trouble focusing. The intense pressure allows him to laser in and obsess over every angle, often to the point that the deal consumes all his waking hours.

Paradoxically, the ability to hyperfocus is a common trait of ADHD, and it often leads people to miss the diagnosis because it doesn't seem to fit. New patients will tell me that they can't possibly have ADHD since they often get completely absorbed in what they're reading or doing. But the glitch in the attention system isn't strictly a deficit—it's more of an inability to direct attention or to focus on command. I tell my patients a more helpful way to think of ADHD is as an attention *variability* disorder; the deficit is one of consistency.

Sam gets this. He schedules important work and meetings early in the day, when he can still feel the calming effects of his morning run, knowing that he gets progressively more scattered as the day goes on. As for returning phone calls, if it weren't for his secretary, it would never happen. He still struggles with most of the core behavioral symptoms that branded him a problem child back in school, but he's figured out how to manage his ADHD and harness his hyperactivity to some extent, even using it to his advantage. By recognizing his difficulties, he's been able to arrange his day and his life in such a way that he can be successful.

MASS DISTRACTION

I have coauthored three books about ADHD with my friend and colleague Ned Hallowell. The first, *Driven to Distraction*, was published in 1994, and as it grew into a bestseller and the broader public became familiar with the general outlines of ADHD, along came the biggest cultural paradigm shift of the century—the World Wide Web. The Internet's endless stream of distractions would challenge anyone's attention span.

It's easy to get distracted in today's world. It's become so full of information, noise, and interruptions that all of us feel overwhelmed and unfocused at times. The amount of data in the world is doubling every few years, but our attention system, like the rest of the brain, was built to make sense of the surrounding environment as it existed ten thousand years ago. In our cybercentric world, however, we quickly learn to expect things to happen immediately, and if they don't, we're easily frustrated. If our cell phone doesn't ring or our e-mail Inbox sits empty for more than an hour, we wonder what's going on. Who has the time or patience to sequence or plan or think things through and evaluate consequences? Why bother, when we can click to the next? It's no wonder exercise gets pushed to the bottom of the priority list—it requires planning and work.

Experts estimate that just over 4 percent of American adults—that's thirteen million people—have ADHD, which is not to say that the remaining 96 percent of the population is completely free of attention problems. To a certain degree, everyone suffers from fleeting attention. And as I've mentioned, there are varying degrees of severity for many mental health disorders—shadow syndromes, which are personality traits that don't necessarily meet the full checklist of symptoms doctors rely on to make diagnoses. People with shadows of ADHD might have constant problems with romantic relationships, for example. Or they might succeed in intense, high-energy fields. Or both. They often become entrepreneurs, bond traders, salespeople, emergency room doctors, firefighters, trial lawyers, movie moguls, or advertising executives. These are jobs in which the tendency toward hyperactivity, nonlinear thinking, and risk-taking can lead to great achievements. Sporadic attention can be a real strength in a frenetic setting. People with shadows of ADHD might retain problems with organization, forgetfulness, and personal relationships, but they can get it together when the pressure is on.

There are still those who think the way Sam's teacher's did. Because all of us suffer from lapses in attention, it's easy to assume that all it takes to concentrate is a little effort. I still meet people who believe ADHD is just a matter of laziness or bad child rearing or stupidity or willfulness or hooliganism. Ironically, such skepticism has roots in the medical community itself, which for decades believed that kids magically grew out of ADHD in adolescence. Some of the work I'm most proud of is challenging this conventional wisdom and showing that ADHD exists in adults.

Now, ADHD is the most-studied disorder in medicine, and it's clearly not an attitude problem. Otherwise it wouldn't run in families. But according to a study of nearly two thousand Australian identical twins, if one twin has ADHD, there is a 91 percent chance the other one will have it too.

The landmark study proving that ADHD stems from a biological irregularity was published by Alan Zametkin and his colleagues from the National Institute of Mental Health in 1990. Using PET scans to measure brain activity, the study showed that during an attention test, the brains of adults with ADHD work differently from those without ADHD. Zametkin and his colleagues found that the group with ADHD showed 10 percent less brain activity than the control group, and the largest deficit was within the prefrontal cortex, which has a firm hand in regulating behavior. It's also prone to positive reinforcement through exercise.

SIGNS OF TROUBLE

The phrase "attention-deficit disorder" didn't even exist until it was introduced in the third edition of the *Diagnostic and Statistical*

Manual in 1980. Since then, we've debated whether to establish separate diagnoses for the two primary categories of symptoms—inattention and hyperactivity. Inattention is always part of the disorder, and sometimes hyperactivity is present as well. Hyperactivity is more common in children than adults and especially—though not exclusively—boys, and for years kids with the rambunctious brand of ADHD were the only cases diagnosed. Nobody connected the behavior of hyperactive kids with that of the daydreamers. But the treatment is the same in either case, and now we call the disorder ADHD, regardless of whether hyperactivity is part of the picture.

The hyperactive kids are the ones you can't miss: they're the Dennis the Menace characters, bouncing off the walls or fidgeting in their seats, constantly moving—picking at themselves, shaking their legs, doodling, fiddling. Because of their impatience, they intrude and interrupt, blurting things out without thinking. They feel as though they're always racing, and they finish our sentences because they assume they know what we're going to say or they're bored by it. In general, they have a hard time staying on task. They can't stand playing by themselves and often adopt the role of class clown if they're having trouble with school. Many of them are socially adept, although they can also act awkwardly because they miss social cues. And, like Sam, they begin to hear early on that they're screwups. But with many of these kids, it's plain as day that they need to be moving, and they end up playing sports and doing well. Impulsivity fits in here as a subset of hyperactivity. Children and adults can automatically overreact, negatively or positively, which makes them passionate and quick to anger. Road rage is essentially a temper tantrum, and a red flag for particularly hyperactive forms of ADHD. Just making it through traffic to my office is a trial for some of my patients: "I wish I had howitzers in my headlights!" one woman told me. "I'd blast everyone out of the way!" Impatience feeds this response too. People with ADHD will do anything not to stand in line and can explode if they're made to wait.

Inattention, or distractibility, is the constant among ADHD symptoms. One couple came in to see me because the wife's inability to pay attention was undermining their relationship. Although she was a whiz at running an intensive care unit—she thrived on the action—she couldn't pay attention to her own family. Even as the husband related this in my office, he suddenly said, "Look!" And sure enough, she was staring out the window. ADHD people go off topic and forget ideas and goals and things. One of the classic signs is the pirouette: stepping out the door, the ADHD patient will spin around and go back upstairs to get something she forgot. Everyone does this, of course, but for some of my patients it's a daily occurrence. The one day the ADHD student does do his homework, he leaves it at home.

The ADHD brain faces a monumental challenge in initiating a task, and it is a master procrastinator. Its owner will sit down to do something she really wants to do and then clean her desk instead. The attention-deficit patient often can't complete something until the Sword of Damocles is over her head. She has a terrible time organizing things, so her room and office are messy. And she has a love-hate relationship with structure. My patient Sam wasn't rebelling against authority per se—he was acting out of frustration from his inability to navigate structure.

Paradoxically, one of the best treatment strategies for ADHD involves establishing extremely rigid structure. Over the years, I've heard countless parents offer the same observation about their ADHD children: *Johnny is so much better when he's doing tae kwon do.* He wasn't doing his homework, and he was angry, difficult, and problematic; now his best qualities have come out.

You could substitute any of the martial arts here or any highly structured form of exercise such as ballet, figure skating, or gymnastics. Less traditional sports, such as rock climbing, mountain biking, whitewater paddling, and—sorry to tell you, Mom—skateboarding, are also effective in the sense that they require complex movements in the midst of heavy exertion. The combination of challenging the brain and the body has a greater positive impact than aerobic exercise alone. One small study from a graduate student at Hofstra University tested this fact. He found that ADHD boys age eight to eleven participating in martial arts twice a week improved their behavior and performance on a number of measures compared to those on a typical aerobic exercise program (both kinds of exercise led to dramatic improvement over nonactive controls). The kids involved in martial arts finished more of their homework, were better prepared for class, improved their grades, broke fewer rules, and jumped out of their seats less often. In short, they were better able to stay on task.

The technical movements inherent in any of these sports activate a vast array of brain areas that control balance, timing, sequencing, evaluating consequences, switching, error correction, fine motor adjustments, inhibition, and, of course, intense focus and concentration. In the extreme, playing these kinds of sports is a matter of survival—avoiding getting karate chopped, or breaking your neck on the balance beam, or drowning in a swirling pool of whitewater—and thus taps into the focusing power of the fight-or-flight response. When the mind is on high alert, there is plenty of motivation to learn the skills necessary for these activities. As far as the brain is concerned, it's do or die. And, of course, most of the time we will be in the aerobic range, which boosts our cognitive abilities and makes it easier to absorb new moves and strategies.

COMING THROUGH, LOUD AND UNCLEAR

The attention system doesn't claim a central address in the brain. Rather, it's a diffuse web of reciprocal pathways that begins at the locus coeruleus, the arousal center, a part of the brain stem, and sends signals throughout the brain to wake it up and cue our attention. The network engages such areas as the reward center, the limbic system, and the cortex; more recently scientists have included the cerebellum, which governs balance and fluidity. It turns out that there's a lot of overlap between attention, consciousness, and movement.

The attention circuits are jointly regulated by the neurotransmitters norepinephrine and dopamine, which are so similar on a molecular level that they can plug into each others' receptors. These are the chemicals targeted by ADHD medications. And of the many genes correlated with the disorder, scientists focus on the ones that regulate these two neurotransmitters. Broadly speaking, the problem for people with ADHD is that their attention system is patchy; they describe it as discontinuous, fragmented, and uncoordinated—problems that can stem from a dysfunction with either of these neurotransmitters or in any one of the brain areas in the system, which helps explain how one disorder can have so many different faces. For instance, the locus coeruleus serves as the on-off switch for sleep and thus is closely tied to circadian rhythms. One of the common symptoms in people with ADHD is abnormal sleep patterns: they often have problems going to sleep or staying asleep, and they suffer sleep disturbances such as sleepwalking or sleeptalking and nightmares. Early theories about hyperactivity proposed that arousal was the chief problem, that children "bouncing off the walls" were essentially trying to keep themselves alert. But the locus coeruleus, which busily produces norepinephrine in the depths of the brain stem, is merely the first opportunity for error. Norepinephrine-carrying axons extending from there, along with dopamine-equipped axons from the ventral tegmental area (VTA), plug into neurons in the amygdala.

As I mentioned in chapter 3, the amygdala is responsible for assigning emotional intensity to incoming stimuli before we're conscious of it, and then sends it along for higher processing. In the context of ADHD, the amygdala determines the "noticeableness" of things. An unregulated amygdala is what feeds the tantrums or blind aggression in patients with ADHD, and their oversensitivity to excitement can lead to panic attacks. Sometimes excitability is a positive — people with ADHD can get so enthusiastic about something that they energize a roomful of people. (Holding the attention of others is no problem for those with ADHD.)

Dopamine also carries signals to the nucleus accumbens, or reward center, which is where Ritalin, amphetamine-dextroamphetamine (Adderall), and the active agents of other stimulants — from coffee to chocolate to cocaine — end up. The reward center needs to be sufficiently activated before it will carry out its important duty of telling the prefrontal cortex that something is worth paying attention to. It engages the prioritizing aspect of executive function, and this is a central component of motivation. Essentially, the brain won't do much unless the reward center is responsive. Laboratory studies have shown that monkeys with lesions in the nucleus accumbens cannot sustain attention and thus can't muster the motivation to perform tasks that don't carry immediate rewards. The same is true of people with ADHD, who favor immediate gratification over more mundane tasks that will help them down the road, like studying for a test that will help them get into college. I call them prisoners of the present. They can't maintain focus on a long-term goal, and so it seems as though they lack drive.

The prefrontal cortex bears responsibility for ADHD too. We can think of inattention in general as an inability to inhibit interest in unimportant stimuli and motor impulses. In other words, we can't stop paying attention to what we shouldn't be paying attention to. The prefrontal cortex is also the home of working memory, which sustains attention during a delay for a reward, and holds multiple issues in the mind at once. If working memory is impaired, we can't stay on task or work toward a long-term goal because we can't keep an idea in mind long enough to operate on it or to ponder, process, sequence, plan, rehearse, and evaluate consequences. Working memory, which is like our random-access memory (RAM), can be considered the backbone of all the executive functions. A failure of working memory is also why people with ADHD are terrible at keeping track of time and thus prone to procrastination. They literally forget to worry about the passing time, so they never get started on the task at hand. An ADHD sufferer who is on the verge of losing her job because she's always late to work might go for the cereal box in the morning, decide the cupboard needs to be rearranged, and forget that she has to be out the door at a certain time. Then, when she remembers, panic sets in.

ATTENTION, ALL CONTROL UNITS

It's not simply a matter of whether the signals get through to capture our attention, but how fluidly that information travels. This is where the attention system ties in with movement and thus exercise: the areas of the brain that control physical movement also coordinate the flow of information.

The cerebellum is a primitive part of the brain that for decades was assumed to be involved only with governing and refining movement. When we learn how to do something physical, whether it's a karate kick or snapping our fingers, the cerebellum is hard at work. The cerebellum takes up just 10 percent of the brain's volume, but it contains half of our neurons, which means it's a densely packed area constantly buzzing with activity. But it keeps rhythm for more than just motor movements: it regulates certain brain systems so they run smoothly, updating and managing the flow of information to keep it moving seamlessly. In patients with ADHD, parts of the cerebellum are smaller in volume and don't function properly, so it makes sense that this could cause disjointed attention.

The cerebellum sends information to the prefrontal and motor cortices—the centers for thinking and movement—but along the route is an important cluster of nerve cells called the basal ganglia, which acts as a sort of automatic transmission, subconsciously shifting attentional resources as the cortex demands. It's modulated by dopamine signals stemming from the substantia nigra. Dopamine works like transmission fluid: if there's not enough, as is the case in people with ADHD, attention can't easily be shifted or can only be shifted all the way into high gear.

A lot of what scientists know about the basal ganglia comes from research into Parkinson's disease, which is caused by a depletion of dopamine in this area. The disease wreaks havoc with a patient's ability to coordinate not only motor movements but also complex cognitive tasks. In the early stages of Parkinson's, these malfunctions show up as adult-onset ADHD.

The parallel is important because, based on a number of strong studies, neurologists are now recommending daily exercise in the early stages of Parkinson's disease to stave off symptoms. Scientists induced Parkinson's in rats by killing the dopamine cells in their basal ganglia, and then forced half of them to run on a treadmill twice a day in the ten days following the "onset" of the disease. Incredibly, the runners' dopamine levels stayed within normal ranges and their motor skills didn't deteriorate. In one study on people with Parkinson's, intensive activity improved motor ability as well as mood, and the positive effects lasted for at least six weeks after they stopped exercising.

What I find so compelling is the strong relationship between movement and attention. They share overlapping pathways, which is probably why activities like martial arts work well for ADHD kids—they have to pay attention while learning new movements, which engages and trains both systems.

A controversial treatment for dyslexia—which occurs in about 30 percent of ADHD patients—relies entirely on physical movements to train the cerebellum. Dyslexia, dyspraxia, and attention treatment (DDAT) is based on the theory that a disruption in the brain's ability to coordinate movement might be responsible for eye-tracking problems and thus difficulties in learning to read and write. Researchers also know that most children with dyslexia perform worse than average on tests of cerebellar function. DDAT involves practicing a collection of fairly simple motor-skills drills twice a day for ten minutes. In 2003 British researchers tested the effectiveness of DDAT on thirty-five children with dyslexia and declared the results "astounding." Compared to no treatment, the students who followed the DDAT regimen for six months showed a significant improvement in reading and writing fluency, eye movement, cognitive skills, and physical measures such as dexterity and balance.

My friend and colleague Ned Hallowell uses this method (among many others) at his ADHD treatment center and has seen the positive effects in his own son. And prominent scientists at Columbia University College of Physicians and Surgeons are just embarking on a large study assessing the usefulness of the DDAT method as a treatment for ADHD.

Pharmacological studies have shown that ADHD drugs help normalize the activity of the cerebellum, as well as the corpus striatum, so it's clear these areas are important to attention as well as movement. Perhaps by training our brain's movement centers to improve its higher functions, we can bring about a day when we're not as dependent on medication.

EARLY CLUES

I have never gotten my taxes in before October. Every year begins the same way, with me resolving to beat the taxman's deadline. In early January I neatly gather all my documents for my accountant. Then, inevitably, I'll find that a monthly statement has gone missing. I need to call my credit card company for a copy, which seems simple enough, but it kills my enthusiasm. The detail of tracking down the missing document, or buying those little white tabs for labeling files, will gnaw at the back of my mind for months. But the momentum is gone and along with it my motivation.

As a child, thankfully, I had strict nuns for taskmasters, and when I wasn't at school I was outside running full tilt in one sport or another. Still, my room was a disaster; I forgot things constantly; and my tennis coach claimed that I was the most consistently inconsistent player he'd ever seen.

I have ADHD, obviously, but I never knew it; the term didn't exist when I was a kid. To the extent that anyone bothered with attention deficit, it was called hyperactivity.

As a doctor, I didn't come across the disorder until I was teaching at Massachusetts Mental Health Center in the early 1980s. My residents presented a twenty-two-year-old patient who had been in and out of the hospital for bouts of violent behavior. He mentioned that he had been on Ritalin as a hyperactive teenager but had long since been taken off the medicine. It was believed that kids simply grew out of their hyperactivity after adolescence and that it was dangerous to keep them on stimulants into adulthood, for fear they'd become addicted. I suggested we try the Ritalin again, and it really toned down his violent outbursts. He was so relieved; he said he'd forgotten that he could feel calm and focused.

Around the same time, I was immersed in studying severe aggression—researching, treating, and writing about what made patients of all sorts so violent. I stumbled across a study by Frank Elliott, then chairman of the neurology department at the University of Pennsylvania. Within a large population of prisoners, he discovered that more than 80 percent of them had had serious learning problems as children.

I started to dig into the school histories of my aggression patients, and common stories emerged. It was clear that they shared a lifelong difficulty inhibiting their thoughts, behavior, and actions. Many of them hated authority, had low self-esteem borne of chronic failure, and were driven by impulse. They acquainted themselves with trouble early in life and never managed to tap into the positive aspects of their personalities. A lot of them had gotten addicted to drugs as teenagers. These tendencies could easily combine with a hair-trigger response to frustration and set off violent episodes. It started to click that such destructive behavior might be rooted in the attention system.

I began looking at my outpatients through the lens of attention. What I saw was that some people with chronic problems such as depression, anxiety, substance abuse, and anger also seemed to share an underlying condition of the attention system, which was easy to miss when it wasn't wrapped in hyperactivity. I began treating them with ADHD medicine, and I saw great improvement. As I discussed my ideas with colleagues, it became clear that there were milder forms of attention deficit that didn't necessarily land someone in prison or the hospital or the unemployment line. Once we looked past the stigma, my friend Ned and I recognized the symptoms in ourselves.

The first paper I wrote on adult ADHD was soundly rejected, based on the criticism that I must be misdiagnosing some form of underlying depression or anxiety or that I was trying to introduce a new disorder. But I knew we were onto something in 1989 when Ned and I gave our first lecture on the subject, at a small conference in Cambridge, Massachusetts, for an organization founded by parents of kids with ADHD. The title of our talk was simply "Adults with ADD" (we didn't call it ADHD back then). After our presentation to a roomful of two hundred people, we figured we'd stick around for fifteen minutes or so fielding questions. We were there for *four hours*. Clustered around the microphone in the aisle, one person after another told bits of their personal stories and asked what they meant. Many of them had the same disorder as their children, and they knew it.

So did a fellow psychiatry professor who came in for treatment after overhearing me at a party one night discuss a case study. "I think you described me," he said, and launched into a highly intellectual rendition of his own history. Charles, I'll call him, was the classic absentminded professor, wearing glasses and unkempt tweed, and he knew a lot more about psychiatry than I did at that point—I'd read several of his books!

The twist to Charles's story is that he had been a marathon runner who had blown out his knee and become depressed when he was forced to set aside his passion. That's also when he noticed the symptoms of what we would agree was ADHD. He explained that he would have a tantrum if his girlfriend interrupted his writing, or yank the phone out of the wall if it rang while he was trying to concentrate. He was slipping out of touch with his friends. He fit the profile, and we decided to put him on ADHD medication, which helped.

He was already on antidepressants when he first came to see me, but once he finished physical therapy and started training again, he dropped them because he felt so much better. As he closed in on his old fitness level, he became convinced that the ADHD medication was holding back his performance. Charles knew his mile times down to the second, and he was ten seconds slower than he used to be.

He decided to try a few days without the ADHD medication, and he found that as long as he was training, he could focus. Looking back on it, we recognized that his attention hadn't hampered him before because he'd always been a serious runner. Without a steady diet of exercise during his injury, he'd been unable to control his attention the way he needed to. Clearly, exercise had a powerful effect, and that was big news to me.

FOCUS ON EXERCISE

Around the time Charles came in, I began to see a number of other intelligent, high-functioning professionals who had ADHD and were able to compensate for it. They didn't fit the stereotype in the literature. Nobody had ever talked about successful adults with attention disorders until Ned and I included their case studies in *Driven to Distraction*. Several of these patients had discovered on their own that they could use exercise as a way of self-medicating to allow them to be more productive. I remember one in particular who is now managing a billion-dollar hedge fund: he takes a stimulant in the morning and plays squash every day at lunch, right about the time the medicine wears off.

Most people instinctively know that exercise burns off energy. And any teacher who has ever dealt with a hyperactive child will tell you that kids are much calmer after recess. Being calmer and more focused is one of the happy consequences of the Zero Hour program in Naperville that I discussed in chapter 1.

School is an excruciating environment for a child with ADHD, given the need to sit still, face forward, and listen intently to a teacher for the better part of an hour. It's impossible for some, and it's the reason for a lot of disruptive behavior among schoolchildren. I got a dramatic reminder of this about ten years ago on a trip to the San Carlos Apache Indian Reservation in Arizona. As part of the tribe's effort to tackle its community's health issues, I was invited to talk about ADHD to doctors, medical staff, parents, and teachers. ADHD is a huge but largely undiagnosed issue for the reservation kids because the incidence of the disorder among Apaches seems to be much higher than for the general population. As I outlined the symptoms and treatments to a group of middle school teachers one afternoon, several of them remarked that all of their kids had trouble sitting still. I asked about recess and was told that the kids have three a day. "If it rains, and they can't go outside," one teacher piped up, "we bus the kids home. Otherwise, we can't handle them."

Incredibly, there are few studies that provide good statistics on the prevalence of ADHD. One of the best comes from the Mayo Clinic. Researchers tracked all of the children born in Rochester, Minnesota, between 1976 and 1982, and followed up with those who stayed in the community until they were five years old. In all, this included 5,718 kids. The study reported that at age nineteen, at least 7.4 percent had ADHD, and suggested that the prevelance might be as high as 16 percent. Other studies suggest that approximately 40 percent of children with ADHD do "grow out" of it, and when it does persist into adulthood, symptoms of hyperactivity often subside. It's no coincidence that the prefrontal cortex, which is responsible for inhibiting impulses, doesn't develop fully until we're in our early twenties. This is the biology of maturity.

ENGAGE THE BRAIN

Given the leading role of dopamine and norepinephrine in regulating the attention system, the broad scientific explanation for how exercise tempers ADHD is by increasing these neurotransmitters. And it does so immediately. With regular exercise, we can raise the baseline levels of dopamine and norepinephrine by spurring the growth of new receptors in certain brain areas.

In the brain stem, balancing norepinephrine in the arousal center also helps. "Chronic exercise improves the *tone* of the locus coeruleus," says Amelia Russo-Neustadt, a neuroscientist and psychiatrist at California State University. The result is that we're less prone to startle or to react out of proportion to any given situation. And we feel less irritable.

Similarly, I think of exercise as administering the transmission fluid for the basal ganglia, which, again, is responsible for the smooth shifting of the attention system. This area is a key binding site for Ritalin, and brain scans show it to be abnormal in children with ADHD. Exercise increases dopamine levels in the rat equivalent of this area by creating new dopamine receptors.

One group of researchers, including the University of Georgia's Rodney Dishman, examined the effects of exercise in ADHD kids by using motor function tests that provide indirect measures of dopamine activity. The results threw Dishman for a loop because boys and girls responded differently. In boys, rigorous exercise improved their ability to stare straight ahead and stick out their tongue, for example, indicating better motor reflex inhibition, which is the missing ingredient in hyperactivity. Girls didn't show this improvement, which may be because of the lower incidence of hyperactivity in girls. Boys and girls both improved by another measure related to the sensitivity of dopamine synapses, although boys fared better after maximal exercise and girls after submaximal exercise (65 percent to 75 percent of maximum heart rate, respectively).

An overactive cerebellum also contributes to fidgetiness in ADHD kids, and recent studies have shown that ADHD drugs that elevate dopamine and norepinephrine bring this area back in balance. Exercise also increases norepinephrine. And the more complex the exercise, the better. Rats don't do judo, but scientists have looked at the neurochemical changes in their brains after periods of acrobatic exercise, the closest parallel to martial arts. Compared to rats running on a treadmill, their cohorts who practiced complex motor skills improved levels of brain-derived neurotrophic factor (BDNF) more dramatically, which suggests that growth is happening in the cerebellum.

In the limbic system, as I've explained, exercise helps regulate the amygdala, which in the context of ADHD blunts the hair-trigger responsiveness a lot of patients experience. It evens out the reaction to a new stimulus, so we don't go overboard and scream at another driver in a fit of road rage, for example.

To the extent that ADHD is a lack of control—of impulses and attention—the performance of the prefrontal cortex is critical. The seminal 2006 study from Arthur Kramer of the University of Illinois used MRI scans to show that walking as few as three days a week for six months increased the volume of the prefrontal cortex in older adults. And when he tested aspects of their executive function, they showed improvement in working memory, smoothly switching between tasks and screening out irrelevant stimuli. Kramer wasn't on the trail of ADHD, but his findings illustrate another way exercise might help.

Everyone agrees that exercise boosts levels of dopamine and norepinephrine. And one of the intracellular effects of these neurotransmitters, according to Yale University neurobiologist Amy Arnsten, is an improvement in the prefrontal cortex's signal-to-noise ratio. She has found that norepinephrine boosts the signal quality of synaptic transmission, while dopamine decreases the noise, or static of undirected neuron chatter, by preventing the receiving cell from processing irrelevant signals.

Arnsten also suggests that levels of the attention neurotransmitters follow an upside-down U pattern, meaning that increasing them helps to a point, after which there's a negative effect. As with every other part of the brain, the neurological soup needs to be at optimum levels. Exercise is the best recipe.

A CLASSIC CASE

If you were to run into Jackson, my former patient, you would meet a compact twenty-one-year-old in jeans and an untucked shirt who speaks articulately about his plans for the future—a typical American college kid, if not a little smarter. What stands out about him isn't so much where he is today but how far he has come to get here and how he did it. Jackson runs nearly every day, three miles on days that he also lifts weights, six miles on the others. "If I don't do it, it's not like I feel guilty," he says. "It's that I feel like I've missed something in my day, and I *want* to go do it. Because I figured out that while I'm exercising I don't have trouble concentrating on anything."

Jackson was fifteen when he first came to see me, for anxiety exacerbated by his ADHD. His tendency to procrastinate invariably put him in impossible situations, and though he prided himself on the craftiness with which he manipulated teachers and academic deadlines, the constant conning took a toll on his nerves. By the end of high school, he had dug himself into a hole so deep that even he wasn't sure he could hustle his way out of it. His future came down to one question on one math test that he postponed until just before graduation. "I dragged it out for so many days that I didn't actually know if I was going to graduate," he recalls. "I was out there in a cap and a gown and didn't know if they were going to say my name." He pauses, and then adds, "I felt dumb."

Jackson was diagnosed with ADHD early on, after his third grade teacher picked up on his disruptive behavior and inability to complete class work. He began taking Ritalin and stayed on some form of stimulant throughout his school years. He was smart, but he had a lot of trouble with school. As a day student at a top-ranked private academy, he simply had more work than he could get through. Sleep became rare, and when it did come he would often wake up with a stomachache, dreading the drive to school. After having a panic attack, he withdrew from the school — despite a B average earned primarily by acing tests — and transferred to a public high school. Unlike some ADHD kids, Jackson was very social, founding after-school clubs and serving as a peer counselor for troubled kids — he figured he had a pretty good handle on psychology after everything he'd learned through his own troubles.

All the extracurricular activity served as a foil for what was turning into severe anxiety and depression, and at one point I had him taking Adderall, paroxetine (Paxil), and clonazepam (Klonopin), a long-acting anxiety drug. Academically, the material was easy enough, but the homework was such a source of stress that either he didn't do it or he rushed through it between classes. He had convinced himself that he was smart enough to pull off high school without really doing the work. He says he felt like a "secret agent man," sneaking around, subverting the attendance rules, and dodging teachers to finish assignments and then feigning innocence. "I thought I was so cool," he says. "My crowning achievement was in this history class that I actually really liked. I didn't do this huge paper, but I managed somehow to trick the teacher into thinking I had—and that I got an A on it. I never handed it in."

They did call Jackson's name at graduation. He squeaked through with a 1.8 GPA, far too low to go to the college he hoped to attend, despite family connections. A small junior college accepted him, though, and that was just fine. The triumph of completing school along with the comfort of having a destination the next fall put him on top of the world. In fact, he felt so good that summer that he decided to go off his medication—all of it. (Needless to say, I was not in the loop at the time.) It was the first time since grade school that he'd gone unmedicated for more than a day or two. "I noticed that a lot of the small things that bothered me went away," he reports. And some not-so-small things: for the first time in his life, he was able to stick to normal sleep patterns, and his anxiety subsided. He figured he was simply feeling great because he had made it through school, but then when he took some ADHD medicine to sit for an English placement exam for college, the irritating side effects returned. After the test, he shelved the meds.

The turning point, however, happened in Spain that summer on a trip with his girlfriend. Walking around shirtless on the beach with all the "Spanish dudes," he was inspired to do something about his Buddha belly. "I just started to run," he says. "And I started feeling great. Part of that, I'm sure, is that I was on vacation in Spain. Everything was great in my life, and I was going to this college that wasn't that hard, so I'm like, Maybe I can just do this! I went to college that fall and I never struggled for a second."

Jackson's story appeals to me partly because he got into exercise for his body image but stuck with it for the therapeutic effect. At first, all the running didn't make a dent in his physique (thanks to pizza and beer), but he stuck with it because it helped him focus. In his first semester at the junior college, he earned a 3.9 GPA, and after a year he was accepted as a transfer student at the college he had originally wanted to attend. It's a small, competitive New England institution, where he earned a 3.5 GPA as a sophomore. His major? Psychology.

He's clearly tuned in to his own state of mind. If he falls off his exercise regimen, his concentration wavers. "I can definitely tell when I don't do it," he says. "It got to the point during midterms that I had no time, but I'm like, You know what? I have to go out and run and clear my head. I *have* to do this."

He knows how it makes him feel, and that knowledge itself keeps him going.

"I always have a million voices in my head all the time," he explains. "When I started exercising, it wasn't that I was just thinking about one thing—because I also have a problem hyperfocusing—but it was like I could concentrate on things that were important to me. Then I started thinking about it, and in general now I really don't have trouble concentrating. And because I'm off the meds, I don't have nearly as many sleep problems. There's never been any question in my mind that exercise is related, because it's this huge life change that I made. It's just so clear."

TAKE THE INITIATIVE

Not everyone with ADHD will experience the sweeping effect of exercise that Jackson did. And I would never have suggested he abruptly quit taking his medication, especially the antidepressant.

His experience begs the question of whether exercise can replace Ritalin or Adderall or bupropion (Wellbutrin), and for the vast majority of cases I would say the answer is no. At least not in the way James Blumenthal and his colleagues at Duke showed that exercise can stand in for Zoloft in treating depression.

Yet there is something instructive in Jackson's motivation for discontinuing his medication. I think he felt out of control, knowing that he was smart enough to succeed but unable to make it happen. Constant frustration can lead to feelings of demoralization, and in Jackson's case, this fed his depression and anxiety. For him, taking medication exacerbated that feeling, creating a sense of dependency. Conversely, getting into a running routine provided a sense of control over his inner self—his mood, his anxiety, his focus. For the first time in his life, he felt like he could steer his own future. He used running as his medicine.

For most of my patients, I suggest exercise as a tool to help them manage their symptoms along with their medication. The best strategy is to exercise in the morning, and then take the medication about an hour later, which is generally when the immediate focusing effects of exercise begin to wear off. For a number of patients, I find that if they exercise daily, they need a lower dose of stimulant.

I'm talking about taking the lead in your own treatment: the more you know about how ADHD works, and the more you recognize your foibles, the better you can prepare for them. I tell my patients they need to develop militant vigilance in terms of scheduling and structure. If you set up your environment in a certain way, you can corral your attention through your own actions and become more productive. Arrange your day and your surroundings in a way that encourages focus and accomplishment—moving the ball forward rather than letting it ricochet off the walls. I'm not suggesting that getting organized and establishing structure can melt away symptoms, but it can funnel your attention in the right direction. Today many people are using ADHD coaches to help them do this. The external accountability is a powerful way to help you maintain routines such as exercise and to meet your goals.

Jackson establishes structure for himself with his daily runs, and this works on two levels: the regular schedule shapes his time so he doesn't have to think about it, and the exercise itself focuses the brain in all the ways I've mentioned.

It's true that many ADHD kids are more active than their peers — studies show they have less body fat, on average — and I see plenty of adults with ADHD who are already exercising. But they need to be doing more, and on a regular basis. In general, I tell my patients to make every effort to institute a regimen of *daily* exercise — or at least during the five weekdays, when they need to focus at school or work. Dishman's study suggests that submaximal exercise, which would be 65 to 75 percent of maximum heart rate, is more effective with girls, while more vigorous exercise (just below the anaerobic threshold, which I'll explain in chapter 10) works better for boys. We don't really have parallel data for adults, but from what I've seen, it's important to get the heart rate up there — maybe 75 percent of your maximum for twenty or thirty minutes.

For ADHD in particular, the complex, focus-intensive sports such as martial arts and gymnastics are a great way to tax the brain. By engaging every element of the attention system, it holds you rapt. These sports are just more interesting than running on a treadmill, and participation tends to be self-perpetuating — it's easier to stick with it.

I try to do my workout first thing in the morning, both for the structure it affords and to set the right tone for the day. A lot of times, that keeps me going. And once I get into the intensity of therapy sessions, it's easy for me to hyperfocus on each patient. Researchers haven't quantified how long the spike in dopamine and norepinephrine lasts after exercise, but anecdotal evidence suggests an hour or maybe ninety minutes of calm and clarity. I tell people who need medication to take it at the point when the effects of exercise are wearing off, to get the most benefit from both approaches.

The truth is, everyone has a different level of attention deficit, and you'll have to experiment to see what regimen works. My hope is that knowing *how* it works will allow you to find the best solution for you. If you want a minimum, I would say thirty minutes of aerobic exercise. It's not a lot of time, especially considering that it will help you focus enough to make the most of the rest of your day.

7

Addiction

*Reclaiming the Biology
of Self-Control*

AMONG THE THIRTY-FIVE thousand people who ran the New York City Marathon in November of 2006 were sixteen former drug addicts, a number of whom joked openly that they'd spent most of their lives "running from the cops." When they crossed the finish line, the distance they had come was far greater than 26.2 miles. Many of them had been imprisoned, homeless, or otherwise destitute when they checked themselves into Odyssey House, a rehabilitation program in New York that treats about eight hundred residents at half a dozen locations throughout the city.

These are the rare, worst-case examples of what can happen to people who completely lose control of their behavior. And while the lives of those addicted to hard drugs such as crack or heroin or crystal meth look drastically different than the lives of those who use or abuse drugs without being hooked, the same principles apply to their brains. Which is to say that the lessons of Odyssey House apply to anyone who struggles with self-control, including those who think of themselves as having addictive personalities. Scientists are now characterizing behavior such as gambling, compulsive shopping, and even overeating in the same biological terms they use to explain substance abuse. The common denominator is an out-of-control reward system, which some people are born with and some people develop.

Odyssey House has been around since the late 1960s, offering services from counseling to job training, elder care to family reconciliation. In the spring of 2000, an employee named John Tavolacci started taking residents running in Central Park, with the goal of training for a 5K charity run held each fall. "We run in groups with them and talk about what goes into running—the discipline, the structure, the teamwork," says Tavolacci, who is now Odyssey House's chief operating officer. "Addicts usually isolate themselves, but here they motivate one another, and they see what it means to set goals and accomplish them."

Many of his charges start out walking, and their first challenge is to follow the one rule Tavolacci imposes: no smoking. Then they build up to running the 1.58 miles around the Central Park Reservoir. About a hundred or so residents participate in the exercise program, which is called "Run for Your Life," and those who become serious runners stay in treatment about twice as long as the nonactive residents. "It sounds obvious," Tavolacci says, "But the only thing that we know about treatment is the longer a person stays in, the more likely they are to succeed."

Odyssey House has always used a holistic approach to treatment and emphasizes the importance of community. This is crucial, according to Odyssey House director Peter Provet, because addiction is such an all-encompassing disorder, cutting into every aspect of life, from family to mood to work. "The drug, for the addict, becomes everything," Provet says. Take it away and suddenly there is an "empty vessel" at the core of the body and mind.

"What better way to start filling the vessel than exercise," Provet suggests. "I strongly believe that exercise can serve as an antidote *and* as a type of inoculation against addiction," he says. "As an antidote, you're giving the individual an avenue of life experience that most have not had—the goals of exercise, the feeling of exercise, the challenge of exercise, the pleasure and the pain, the accomplishment, the physical well-being, the self-esteem. All that exercise gives us, you're now presenting to the addict as a very compelling option."

Inoculation ranks as equally important, given that most addicts fight a protracted, sometimes lifelong, battle with relapse. And Provet sees exercise as the best form of inoculation. "Exercise is directly *antithetical* to drug-addictive behavior. Because you need lung strength, muscle strength, mental acuity to engage in physical exercise — lots of things that drugs deprive you of. If you're not eating, not caring about your body, letting it waste away, having your mind distorted by being constantly intoxicated, you can't be a serious exerciser. *You can't do it.*"

Neurobiology is just catching up with what twenty years of experience have taught Provet. The way he describes exercise's effect on the addict mirrors what I discussed in chapter 5, about depression. As a treatment, exercise works from the top down in the brain, forcing addicts to adapt to a new stimulus and thereby allowing them to learn and appreciate alternative and healthy scenarios. It's activity-dependent training, and while it may not provide the immediate rush of a snort of cocaine, it instills a more diffuse sense of well-being that, over time, will become a craving in its own right. The inoculation works from the bottom up, physically blunting the urge to act by engaging the more primitive elements of the brain. Exercise builds synaptic detours around the well-worn connections automatically looking for the next fix.

"Not everyone is going to become a marathon runner, but more and more we're going from addict to athlete," says Provet. "Is it for everybody? Probably not. Is it for most people? Probably yes."

UNJUST REWARDS

As with so many discoveries about how the brain works, scientists stumbled on the first clues about addiction by accident. In 1954 psychologist James Olds and a graduate student named Peter Milner, at McGill University in Montreal, were studying behavior by inserting electrodes into the brains of live rats. They wanted to pinpoint an area related to learning, but in one of the animals, the electrode ended up in the wrong spot. The result was even more interesting than what they were looking for: The rat kept returning to the corner of its cage where it received its first jolt. To the researchers' amazement, they found they could steer the rodent like a remote-control toy by doling out bursts of electricity. The next day, the rat sought out the same corner. Clearly, the rat wanted the stimulation, so much so that it would ignore food placed in one corner in order to receive the shock in another.

In the most famous of their experiments, Olds and Milner rigged up a lever so the rat could administer its own brain stimulation. After discovering that pressing the lever delivered a jolt, it pressed it about once every five seconds until the juice was switched off. Then the rat pawed at the lever a few times with no result and promptly fell asleep.

The brain area that Olds and Milner hit with that electrode is closely related to the nucleus accumbens, or reward center, and it has been the focus of addiction research ever since. It's a critical node of the attention system, as I described in the last chapter, and it's also important in addiction. The reward center provides the necessary motivation for the brain to learn behavior that brings us things we like or want or need. All the things people become addicted to—alcohol, caffeine, nicotine, drugs, sex, carbohydrates, gambling, playing video games, shopping, living on the edge—boost the dopamine in the nucleus accumbens. Regardless of the varying psychological effects different drugs have on the mind, they all boost dopamine in the reward center. As an illustration of the power of drugs, consider that while sex increases dopamine levels 50 to 100 percent, cocaine sends dopamine skyrocketing 300 to 800 percent beyond normal levels.

The nucleus accumbens used to be known as the pleasure center, fueling the notion that addicts are essentially looking for a good time. And while pleasure is certainly an initial factor in enticing people to try a drug or their luck at a gaming table, it's not quite right to think of addicts purely as hedonists. Nobody *enjoys* being addicted. Indeed, by studying how dopamine works as the key messenger in the reward system, scientists have drawn a distinction between liking something and wanting it. "*Liking* refers to the actual experience of pleasure, versus the motivational state, which is the willingness to work for rewards," says Terry Robinson, a behavioral neuroscientist at the University of Michigan. "Dopamine is involved in this wanting, but it's not involved in liking."

The reward center is where ADHD and addiction overlap, which explains why both problems undermine motivation, self-control, and memory. It is no coincidence that about half of those with ADHD also struggle with substance abuse of some kind. The implications have changed the way scientists describe addiction.

The pivotal issues seem to be salience and motivation rather than pleasure. In this context, *salience* means something that stands out against the landscape of life, predominating over all other stimuli. Cues for both pleasure and pain send dopamine coursing through the nucleus accumbens to attract our attention so we can take action to survive. For the developing substance abuser, the overload of dopamine has tricked the brain into thinking that paying attention to the drug is a matter of life or death. "Drugs are tapping into the very core systems that have evolved to mediate survival," says Robinson. "They activate the system in ways it was never meant to be activated."

The National Institute on Drug Abuse now defines addiction as a compulsion that persists in spite of negative health and social consequences. Plenty of people use and abuse drugs, but only relatively few become addicts. Why? While dopamine in the reward center creates the initial interest in a drug or behavior and provides the motivation to get it, what makes addiction such a stubborn problem is the structural changes it causes in the brain. Scientists now consider addiction a chronic disease because it wires in a memory that triggers reflexive behavior. The same changes occur regardless of whether the addiction is to drugs or gambling or eating.

Once the reward has the brain's attention, the prefrontal cortex instructs the hippocampus to remember the scenario and sensation in vivid detail. If it's greasy food that you can't resist, the brain links the aroma of Kentucky Fried Chicken to Colonel Sanders's beard and that red and white bucket. Those cues take on salience and get linked together into a web of associations. Each time you drive up to KFC, the synaptic connections linking everything together get stronger and pick up new cues. This is how habits are formed.

Typically, when we learn something, the connections stabilize and the levels of dopamine tail off over time. With addiction, especially drug addiction, dopamine floods the system with each drug use, reinforcing the memory and pushing other stimuli further into the background. Animal studies show that drugs such as cocaine and amphetamine make the dendrites in the nucleus accumbens bloom, thus increasing their synaptic connections. The changes can remain months and maybe even years after the drugs are stopped, which is why it's so easy to relapse. One way to look at addiction is that the brain has learned something too well. These adaptations lead to a vicious cycle in which the basal ganglia goes on autopilot whenever you smell fried chicken, and the prefrontal cortex can't override your actions even though you may know better.

One of the responsibilities of the prefrontal cortex is to assess risk versus reward and to decide whether to inhibit behavior that may cause harm. With addicts, it's not so much that they make bad choices as that they fail to inhibit behavior that has become reflexive. We know from studies in animals and humans that cocaine, for one, damages nerve cells in the prefrontal cortex and even reduces gray matter. And in recent years, imaging studies have shown that the prefrontal cortex doesn't fully develop until we are well into our twenties, which could explain why most people who experiment with drugs and get hooked do so as teenagers or during early adulthood, when their inhibition hasn't fully developed. "They end up with a hypersensitive system that wants drugs, and they make very bad decisions," says Robinson. "It's the worst of both possible worlds."

GETTING BACK ON YOUR FEET

There's nothing like appearing before a judge to hasten the development of a teenager's inhibition. A patient of mine named Rusty might have ended up a drug addict, but the prospect of three years in jail scared him into cleaning up his act, and the exercise routine he developed in the process is why he's still on the right track today.

I began treating Rusty the summer after his sophomore year in high school, a few months after he was hospitalized for attempting suicide. Feeling lonely and outcast, he had washed down a stash of pills with a pint of peach schnapps. He had good test scores—along with bad grades and a history of tantrums—but no friends at all. It was clear to me that he suffered from an attention deficit, combined with fairly severe symptoms of what I call social dyslexia, meaning he didn't know how to talk to people or be relaxed and flexible in conversation. Rusty's strategy to be cool and make friends was to dress in all black and sell marijuana that he grew himself.

I put him on a long-acting stimulant—a drug he couldn't abuse—for ADHD, and he picked up his grades a little and did very well on the SATs the spring of his junior year. Still, whenever he felt bored or at a loss, he would take anything he could get his hands on, from cocaine to cough syrup. Home alone one afternoon his senior year, he had a panic attack from too much cocaine and called 911. An ambulance came right away—along with the police, who found drugs in his room. He was arrested for possession and intent to distribute and spent the night in jail.

A court date was set for four months down the road, and his lawyer and I worked out a treatment plan—each week he had to take two drug tests and attend a meeting at Alcoholics Anonymous and one at Narcotics Anonymous. He knew he had to stay clean at least until his court appearance, but he started to crave cocaine. His lawyer told him he'd probably get the maximum sentence of three years in jail, and he desperately wanted help. Dealing with his cravings was our first order of business, and I told Rusty that exercise could have a huge impact. He didn't like running or sports, and aside from a stint playing soccer as a child, he was essentially physically inert. I had just returned from my first visit to Naperville, and perhaps because of the way he dressed, I thought of a goth girl named Rachel who had really transformed herself by playing Dance Dance Revolution (DDR), the interactive video game in which the player controls the action on screen by dancing on a mat that's connected to the television. The footwork involved is exhausting even to watch, like the tire drill football players practice, except that the game gets faster and faster at each level.

Rusty agreed to try it, and although he felt clumsy at first, he started to enjoy it. Almost immediately, he said, it blunted his cravings. With nothing much to do that summer except worry about whether he'd be going to jail, he used the game to fill his time and to medicate himself. Guarding against boredom is critical because idle time is dangerous for someone fighting a drug habit.

Rusty got to the point where he was playing DDR several hours in the morning and at night, every day. I saw that his energy level and his optimism picked up. I wrote a letter to the judge, and Rusty was put on probation rather than given jail time, on the condition that he would continue drug testing, Narcotics Anonymous, and counseling at college. He took his DDR setup with him and continued doing it every day for a while. Then he joined an intramural soccer team and started going to the gym.

Exercise was a conduit for shifting Rusty's focus to a more productive life. I see exercise as a way of offsetting the feeling of hopelessness and uselessness that a lot of drug users have, and that certainly was a factor with Rusty. The routine and the physical activity gets the brain engaged and the mind moving in a direction other than toward the drug, reprogramming the basal ganglia to wire in an alternative reflexive behavior. Many people retreat to the couch and give up, but being in motion fosters the feeling that you can accomplish something.

A doctor named Gene-Jack Wang, one of the country's foremost addiction researchers and chairman of the medical department at Brookhaven National Laboratory, talks about movement in philosophical terms. "In the Chinese language, a subject is an animal, and an object is a vegetable," he says. "You cannot ask a vegetable to jump from here to there. If you don't move, you are not an animal anymore—you become a vegetable!"

Certainly this is a factor with the marathoners of Odyssey House. But even with a milder case such as Rusty's, doing DDR chased the bleakness out of his view of his future. And while most experiences will pale in comparison to the high of snorting cocaine, the possibility of a rich life can help keep that memory in perspective.

Rusty is now in his sophomore year of college, making good grades, and dating a girl who is also committed to staying sober. He's taken a leadership role in his dorm, and he's gotten into rock climbing as well as playing soccer. He even started scuba diving, a family activity that he'd avoided in the past. After a recent diving vacation, he told me that he is constantly amazed to see how rich and colorful natural life can be.

CRAVING A DOPAMINE FIX

What Rusty finally sees—that he can find pleasure without drugs—is vital to resisting the urge. When you talk to hard-core addicts, you often hear that they feel numb to most things. Naturally satisfying forms of stimulation such as love, food, and social interaction are a bland backdrop to the vivid experience of the drug. The normal course of life doesn't do it—they can't *feel* it.

Some people are simply born this way. A groundbreaking study in 1990 revealed, for instance, that a lot of alcoholics have a gene variation (the D2R2 allele) that robs their reward center of dopamine receptors, lowering levels of the neurotransmitter. Presence of the D2R2 allele doesn't guarantee you'll end up as an addict, but it's more likely. While 25 percent of the general population has the variation, in one study researchers found it in 70 percent of alcoholics who had cirrhosis—presumably the most addicted, since they continued drinking in the face of life-threatening liver damage. In a subsequent study of cocaine addicts, half had the D2R2 allele, and 80 percent of those who also abused other drugs had it. Results tell a similar story with gamblers and the morbidly obese: about half display the gene variation, but when we factor in other addictive behavior, it's more like 80 percent. Researchers named this problem reward-deficiency syndrome, and the media declared that scientists had found the "alcoholic gene."

Unfortunately, it's not that simple. Without question, if the reward center isn't receiving enough input, you're genetically predisposed to be constantly craving, relentlessly searching for a way to compensate for the deficit. Reward deficiency also undermines the attention and stress systems: when dopamine is out of balance, the amygdala gets involved because it thinks survival is at risk, and that intensifies the pursuit of bringing the brain to equilibrium. This relates back to why so many people with ADHD are seen as "stress junkies"—cortisol quickly boosts dopamine to improve attention. You can see how this nagging feeling — people describe it as a hollowness inside — could leave a person vulnerable to addictive behavior, from taking drugs to gorging on chocolate to playing video games forty hours a week.

But just because you have reward-deficiency syndrome doesn't mean you're destined for Odyssey House. There are hundreds, if not thousands, of factors that influence addiction, and the drive to find something new and exciting can just as easily turn people into bold explorers, iconoclastic artists, or maverick entrepreneurs or send them down any number of paths where pushing conventional boundaries and seeing the world differently are highly valued commodities.

Not surprisingly, athletes in high-risk sports like skydiving display less inhibition and more thrill-seeking behavior than, say, rowers. A recent study from Holland also showed that, like hard-core addicts, many skydivers don't experience pleasure from typical daily life. Both skydivers and addicts have a higher-than-normal threshold for excitement, but is that the cause or the result of the dopamine-boosting behavior? Other research shows that drugs such as cocaine damage D2 receptors, the slots the neurotransmitter plugs into to signal salience. If you continually subject your brain to an overload of dopamine, the number of receptors will dwindle. So regardless of what your brain looked like when you were born, the more drugs you take, the more drugs you'll need to feel the same rush. The same is true of people of people who overeat: "You need more, more, more to make you feel good," says Brookhaven's Gene-Jack Wang.

FIGHT THE URGE, SHAKE THE HABIT

A study in London in 2004 showed that even ten minutes of exercise could blunt an alcoholic's craving. The researchers divided forty hospitalized patients who had just completed detox into two groups: one was assigned to stationary cycling at moderate intensity, the other, light intensity. The next day they switched the groups and found that intense exercise significantly reduced the urge for a drink. This is what happened with my patient Susan, from chapter 3, who used her jump rope to fend off the stress-induced urge to drink wine in the middle of the day.

The biology of stress ties in with addiction in that withdrawal puts the body in survival mode. If you suddenly quit drinking, for instance, you're turning off the dopamine spigot and the hypothalamic-pituitary-adrenal axis gets thrown out of balance. The intense unpleasantness of withdrawal lasts for only a few days, but your system remains sensitive for much longer. If you're in this delicate state and come under further stress, your brain interprets the situation as an emergency and sends you looking for more alcohol. That's how a problem at work or a fight with a lover can cause a relapse. For someone who's been dependant on drugs and has altered his dopamine system, the most effective solution to a stressful situation—and the only one he knows—is the drug. But exercise is another solution.

In smokers, just five minutes of intense exercise can be beneficial. Nicotine is an oddball among addictive substances as it works as a stimulant and a relaxant at the same time. Exercise fights the urge to smoke because in addition to smoothly increasing dopamine it also lowers anxiety, tension, and stress levels—the physical irritability that makes people so grouchy when they're trying to quit. Exercise can fend off cravings for fifty minutes and double or triple the interval to the next cigarette. And the fact that exercise sharpens thinking comes into play here, because one of the withdrawal symptoms of nicotine is impaired focus. As evidence of this, one study found that there are more workplace accidents during the Great American Smokeout than on any other day of the year. Many of my ADHD patients use cigarettes to help them focus when they have to write or push through a challenging task, and without the nicotine they feel lost.

Some drugs, of course, dull the brain to begin with. A novel study from researchers in Iran recently examined how exercise affects rats on morphine. Their hypothesis was that since exercise influences dopamine and plasticity in the same brain areas involved with addiction and learning, maybe it would counteract the memory loss that goes along with being high. The scientists conditioned the rats by putting them in a dark box in which the floor shocked their feet, and then they did follow-up tests to measure how long it took the rats to move to another box that was harmless but well-lit (rodents prefer darkness).

The rats were divided into four groups: one group ran on a treadmill and received a shot of morphine before each trial; one ran and received a placebo injection of saline; another received morphine but didn't exercise; and a control group received neither exercise nor an injection. Both groups that exercised remembered that the dark box was bad news: they hesitated the longest to enter it and were the quickest to leave when they were shocked. Amazingly, the exercise-and-morphine group performed better than the control group, indicating that exercise offsets the mind-dulling effects of the drug.

In the same study, the researchers found that exercise dramatically reduced withdrawal symptoms in the exercise-and-morphine group when they cut off the drugs—in rats, the signs of withdrawal are identified as "wet dog shakes," writhing, and diarrhea. This fact alone should be enough to convince a recovering addict to lace up his sneakers, and it lends scientific credence to the treatment approach at Odyssey House.

A TALE OF DEPENDENCE

Over the years, I've seen many people with reward-deficiency syndrome. The most dramatic example is a Dutch woman I'll call Zoe, who suffers from severe ADHD and has a tumultuous history of depression, aggression, and a range of substance abuse. Most notably, she was a chronic marijuana smoker for twenty years who believed that self-medicating was the only way she could feel calm and focused.

In reality, she was trying to blot out the frustration and anger of her life. As a child, Zoe told me, she was combative and had severe learning problems. Now forty, she is still prone to tantrums and anxiety. On one occasion, when she was flying to Boston for a visit, she erupted in a panic attack and forced the plane to return to Amsterdam.

Zoe spent thirteen years getting through college, which is a long time even for her field of veterinary medicine, partly because she wasn't diagnosed with ADHD until she was twenty-seven. She received a prescription for Ritalin but first had to go to a detox clinic to give up marijuana. "I was smoking ten or twenty joints a day," she recalls. "When I was in there, I was like a wild animal in a cage." She stopped smoking marijuana for about a year, but then she relapsed and soon fell back into her heavy habit of staying high throughout the day (while also taking the Ritalin and antidepressants).

Although she found a high-level job in her profession, Zoe stalled out in the decade after college and in many ways gave up on developing herself. Because she was always so driven to find an immediate reward, she didn't set goals and strategies to move forward in life. Zoe often complained that she felt like life wasn't worth it. Smoking marijuana, she says, kept her from dwelling on the fact that she was unhappy and discontent.

She had always exercised sporadically—cycling, sailing, and horseback riding—but I raised the subject of doing something on a regular basis. I appealed to her knowledge of medicine and explained how exercise could change her brain chemistry and rewire connections in the pathways controlling her mood, aggression, and attention as well as her addiction. After reading several of the studies I've included in this book, she committed to giving daily exercise a try, and she quit smoking marijuana again. "There was no alternative," she says. "I had to do something."

What she did was get an indoor bike trainer of the type used by serious cyclists to hone their balance and stamina—you pedal on free-spinning drums, acutely aware of the possibility that you could slip off and careen across the room. I'm not sure how Zoe settled on this extremely challenging form of exercise, but it has worked out tremendously well. The balance and precision required to ride on the rollers engages the entire attention system, from the motor centers of the cerebellum and basal ganglia to the reward center and prefrontal cortex. "At first I hated to do it because it doesn't bring you anywhere," she says. "Now I'm very handy on it, and it's beneficial because it makes me concentrate as well as exercise. It's exciting because you don't want to fall down."

As if kicking her habit wasn't difficult enough, Zoe's husband left her in the midst of her effort to stay sober, just before Christmas. I was worried, and so was she. "During the winter it gets cold and very dark in Holland," she wrote in an e-mail. "I was so scared that I would get depressed again and go back on pot, but I haven't. The change comes from the difference between feeling like a loser (smoking) and a winner (exercise)."

Zoe's recovery is tenuous, as it is for any long-term drug user. But she's certainly on the right track. She sends me regular updates about trying to break her record distance on the bike trainer, and she's also taken up the jump rope. Here's a snippet from one of her typically buoyant messages: "I just did 10 minutes of rope jumping, heart rate 140, exhausting but I had to do it. This is so GOOD, because in 10 minutes it feels like a half hour biking! Maybe I'll continue this—it's a FAST REWARD!!! It's the exercise I'm craving nowadays."

A NATURAL HIGH

Some would debate whether Zoe was addicted to marijuana, but there's no question she was *dependent* on it. She had all the signs of chemical dependency, including the physical and emotional irritability of withdrawal. Studies in rats show that if they get used to a chronic dose of tetrahydrocannabinol (THC) — the active compound in marijuana — and then are deprived of it, the brain floods the system with corticotropin-releasing factor, which activates the amygdala and thus the entire stress system. The rodents experience shaking, tremors, and twitchy movements that peak about forty-eight hours after the last dose. Indeed, Zoe felt like a rat in a cage when she went through detox: along with the physical symptoms, the shutdown of the dopamine system brought on intense feelings of depression and anxiety. The way exercise blunts the symptoms of withdrawal is by calming the amygdala and boosting dopamine.

Regardless of whether there is such a thing as marijuana addiction, studying THC's effect on the brain has provided new clues about how exercise counteracts addiction of any stripe. To begin with, the feeling that often comes after exercise can serve as a harmless replacement for the drug high. In a recent study in the *British Journal of Sports Medicine,* researcher Arne Dietrich wrote that the way people describe runner's high is "similar to the claims of distorted perception, atypical thought patterns, diminished awareness of one's surroundings, and intensified introspective understanding of one's sense of identity and emotional status made by people who describe drug or trance states."

We've been on the case of the runner's high for three decades, and in the last few years the focus has expanded beyond endorphins to include endocannabinoids, a class of neurotransmitters. Endocannabinoids are to THC as endorphins are to morphine — substances produced in the body that elicit the same effect as a drug. Likewise, they both dull pain.

Scientists discovered the endocannabinoids in the early 1990s after realizing that THC binds to specialized receptors in the brain. These receptors didn't evolve for us to enjoy marijuana, obviously, so there had to be some natural substance the body produces for them. What they found were the neurotransmitters anandamide and 2-arachidonoylglycerol (2-AG). It turns out that marijuana, exercise, and chocolate all activate these same receptors in the brain.

Both of these endocannabinoids are produced in the body and the brain when we exercise. They travel through the bloodstream to activate receptors in the spinal cord, and this blocks pain signals from getting to the brain (not unlike morphine). They also move throughout the reward system and the prefrontal cortex, where they have a direct effect on dopamine. When the endocannabinoid receptors are strongly activated, they produce all the euphoric feelings of marijuana, and, along with endorphins, they act as the body's extra-strength aspirin. Doctors are starting to use anandamide to treat pain syndromes such as chronic fatigue and fibromyalgia, and a number of studies have shown that gradually increasing exercise can relieve the pain and fatigue associated with these syndromes. The link between exercise and these natural pain killers makes perfect sense: they evolved to help us deal with the inevitable pain of straining muscles and joints during the hunt.

Unlike endorphins, endocannabinoids pass easily across the blood-brain barrier, which for some researchers, makes them a more plausible explanation for runner's high. In 2003, a group led by psychologist Philip Sparling of Georgia Tech University showed for the first time that exercise activates the endocannabinoid system. Using fit, male college students either running on treadmills or cycling on stationary bikes for fifty minutes at 70 to 80 percent of their maximum heart rate, the researchers measured how the effort affected blood levels of anandamide. The result? Anandamide nearly doubled.

Runner's high itself is difficult to study because it's so unpredictable — even marathon runners don't experience the feeling every time they train. And why isn't there such a thing as swimmer's high? One intriguing theory is based on the relatively new finding that there are endocannabinoid receptors in the skin that may be activated only by all the pounding and jostling of running. Regardless of whether the particular gauzy delirium of runner's high kicks in, Sparling's work clearly implies that the boost in anandamide is at least one reason we feel relaxed and satisfied after moderately intense exercise. Scientists still debate whether endorphins are involved, however, and it seems likely that the overall effect is some combination of these factors.

HOOKED ON THE GOOD STUFF

If exercise acts like certain drugs in the brain, then you might wonder whether it can also be addicting. I get this question all the time, and the short answer is yes, but don't worry about it. Scientists have tested exercise addiction in rats and found that if they have unrestricted access to a running wheel and are given food only one hour a day, the rats will log about six miles a day and eventually run themselves to death. They don't learn that they need to get all of their nutrition during their one-hour feeding. The more they run, the less they eat, and their calorie intake falls short of the output. They are addicted just as they would be to cocaine. Curiously, the experiment doesn't work if a treadmill is substituted for the running wheel; perhaps there's something about the infinite nature of chasing the next rung that keeps the rats hooked. Whatever the case, the running wheel is a perfect metaphor for addiction.

The danger of getting addicted to exercise applies to a very small segment of the population, most notably girls with anorexia or anyone with a body dysmorphic syndrome, a mental disorder defined by a preoccupation with a perceived deficit in appearance.

They eat less and less, and when they exercise they become light-headed and exhilarated, the high only reinforcing the cycle. They feel great for a short while, and they think they are on their way to looking great. Sadly, their approach will never get them there. But for the vast majority of people, this trap is a remote danger. Even if exercise becomes a dependence—as it could for Zoe, for example—there's little to worry about.

I can think of no better example of somebody with an exercise dependence than ultramarathoner Dean Karnazes, the forty-four-year-old Californian who has appeared on *60 Minutes*, *The Tonight Show*, and countless magazine covers for his mind-bending feat of running fifty marathons in fifty days (in fifty different states). He also ran 350 miles without stopping. Only slightly less impressive to me is that over the past fifteen years, the longest period he has gone without exercising is *three days*. "I had the flu," Karnazes recalls. "I was still sick, but I finally said, Screw it, I need to bust out a run." For starters, his streak says something about the formidable strength of his immune system.

Karnazes was drunk at a bar on his thirtieth birthday when he decided to make a change in his life—at that very moment. He stumbled home, grabbed an old pair of sneakers, and ran thirty miles into the night. He was no alcoholic and has never been into drugs, but the question remains: does this guy have a problem? "Maybe 10 to 20 percent of the time I do think of running as an addiction," he says. "What I really long for is the feeling of bliss or satiety that I get after exercise. It makes me feel complete. I'm most tuned into this when I *can't* exercise. If I'm traveling or I'm in meetings all day, I can feel it pulling at me. I'll think, Why am I ready to implode? I'm crawling out of my skin here! And then I realize my body needs to move. It's almost a feeling of being trapped."

There is no typical week for Karnazes, but he says he averages seventy to ninety miles—three to four hours on most days. In other words, he moves more in a day than most Americans do in a week. This frightens people. It would be easy to paint Karnazes as a freak of nature, and plenty of people have. When you talk to him, however, it seems that despite the huge time demand of training, he leads a balanced life. He held nine-to-five jobs at Fortune 500 companies for more than a decade, and then became president of a natural snack-food company, before recently transitioning into a professional athlete and author (his book, *Ultramarathon Man*, is a bestseller). He has two children, ages eleven and nine, whom he tucks in most every night and drives to and from school every day. He typically wakes up at around three a.m., after four or five hours of sleep, to get in his training before the kids need to go to school.

"I have built my lifestyle around running, so I can support this level of activity," Karnazes says. "Perhaps it's an addiction—I don't know; I've never been through psychoanalysis. I'm just listening to my hardwiring. Luckily, I'm not shooting something in my veins or hopping down to the bar every night after work. Exercise is the ultimate drug, right? What drug always works and doesn't have any unhealthy side effects?"

FILLING THE VESSEL

My patients, Rusty and Zoe are inspiring examples of people who have replaced addiction with exercise, instituting routines in their lives that serve as a healthy alternatives to the full-time pursuit of drugs. As I explained, the addict's brain adapts at every level to focus attention and effort around getting the reward. The brain functions the same way whether the addiction is to alcohol, drugs, food, gambling, or any other addictive substance or behavior. As addiction progresses, there is less and less room for anything else in life.

When an addict quits, what's left is emptiness. In this respect, dealing with addiction is similar to battling feelings of anxiety and depression: getting rid of the problem is only the first step. Once the addiction or the negative emotions are gone, the void needs to be filled with some positive behavior for the change to take root. There can hardly be a better option than physical exercise. After all, this is what we're supposed to be doing—moving in the world.

The fact that exercise counteracts anxiety and depression directly can have a huge impact on any form of addiction, as both of these mood states undermine treatment. A recovering addict who is feeling anxious or hopeless is much more likely to slip in her determination and ability to quit. People are more impulsive when they feel lousy. Both strength training and aerobic exercise decrease symptoms of depression in recovering alcoholics and smokers who have quit. And as I pointed out in chapter 3, the more fit you are, the more resilient you are. If you are flexible in managing stress, you're less likely to reach for that bottle of liquor or bag of chips or pack of cigarettes. Keeping the stress system under control is also important, practically speaking, for ameliorating the physical symptoms of withdrawal, to get through those nightmarish first few days.

Exercise also counteracts more direct toxic effects of addiction on the brain. Researchers looking at fetal alcohol syndrome, for instance, have shown that exposing unborn rats to high levels of alcohol dramatically reduces the birth of new brain cells in the hippocampus. It also disrupts long-term potentiation (LTP), the cellular mechanism of learning and memory. Studies of adult rats exposed to alcohol before birth suggest that they have difficulty learning.

The exciting news on this front is that both exercise and abstinence from alcohol not only stop the damage but also reverse it—increasing neurogenesis and thus regrowing the hippocampus of adult rats. The same holds true even for unborn rats if their mothers are taken off ethanol and allowed to run. In humans, researchers have recently shown that abstinence reverses some of the neuronal damage caused by prenatal exposure to alcohol, and we already know that exercise rebuilds the alcoholic brain by increasing neurogenesis.

One of the connections I see here is between learning and overall mental strength. If the brain is flexible, the mind is stronger, and this gets at a concept known as self-efficacy. It's difficult to measure, but it relates to confidence in our ability to change ourselves. For most addicts, if they stop to consider how they may be destroying their lives, they suddenly feel like they can't handle anything, let alone their self-control over their addiction. Exercise, though, can have a powerful impact on the way an addict feels about himself. If he's engaged in a new pursuit such as exercise, which involves work and commitment, and he's able to follow through and be persistent with it, that sense of self-control spreads to other areas of his life.

A group of Australian researchers recently put this idea to the test. Using twenty-four students as subjects, they measured the effect of a two-month exercise program on self-regulation, which is a slightly different characterization of self-efficacy. Every two weeks, the students were given two psychological tests, and they kept diaries of their daily habits. The results, published in 2006 in the *British Journal of Health and Psychology*, are profound. Aside from improving on the two tests, which measured intellectual inhibition (control), the participants reported that an entire range of behavior related to self-regulation took a turn for the better.

Not only did they steadily increase their visits to the gym, they reported that they smoked less, drank less caffeine and alcohol, ate more healthy food and less junk food, curbed impulse spending and overspending, and lost their tempers less often. They procrastinated less and kept more appointments. And, they didn't leave the dishes in the sink — at least not as often.

The researchers characterized self-regulation as a resource that can be depleted but also recharged like a muscle. Essentially, the more you use this faculty, the stronger it gets. And exercise is by far the best form of self-regulation we have.

REGAIN CONTROL

I wouldn't suggest that you model your routine after that of Dean Karnazes, but if you have a tendency toward addictive behavior it's vital to develop some sort of consistent exercise habit.

How much exercise you need depends, of course, on how severe the habit is. But I would say thirty minutes of vigorous aerobic exercise five days a week is the bare minimum if you want to root out an addiction. To begin, however, it's best if you can do something every day, because the exercise will keep you occupied and focused on something positive. I have seen a lot of people who bury themselves in addiction when they lose their jobs, so if you are unemployed, having exercise in place is essential. And while I often suggest that people exercise in the morning, if your goal is to break a habit such as having a drink every night when you come home, exercising in the evening is probably a better strategy. You can use the aerobic shot for a different kind of buzz.

At the same time, you have to be careful not to overdo it and to find something you'll be able to keep up over the long haul. The patients I've told you about all learned that aerobic exercise provides a strong reward, and they have been able to find that sense of satisfaction in a variety of activities. Rusty couldn't do DDR all the time, so he got into soccer again and picked up rock climbing. Zoe started on the rollers, but as soon as spring hits she's outside, riding her bike through the forest. The more options you have, the more likely you are to be able to continue exercising throughout your life.

If you haven't been in the habit of exercising, it can be helpful to join a gym or hire a personal trainer, because spending the money is a strong motivator. If you have an addiction to food, try a quick walk around the block or a few minutes with a jump rope or even a set of thirty jumping jacks—anything to snap your mind out of the cycle of thinking about the reward.

It might sound painfully obvious to suggest exercise as a way of controlling your eating habits. After all, your weight is the sum of a simple formula—the number of calories you take in minus the number you burn. But it's important to remember that exercise's benefits go well beyond the physical aspect of burning calories. Dopamine produced during exercise will plug into receptors and thus blunt the craving, and over time the activity will produce more D2 receptors and restore balance in the reward system. For someone with a negative body image, shifting the focus from the body to the brain can provide a powerful new sense of motivation.

A lot of people assume that an addict's real problem is just a lack of motivation. On one level, this is true, but what very few people recognize is that motivation is a function of brain signals, and that those signals depend on reliable messengers and intact nerve pathways. When we look at addiction as a neurological malfunction rather than as a moral failure, it suddenly takes on the form of something that can be fixed. It's certainly not an easy task, but it's a lot easier when we use exercise as a tool, one with great versatility. Exercise isn't necessarily a cure, but it's the only treatment I know of that works from the top down as well as from the bottom up, rewiring the brain to circumvent the addictive pattern and curbing the craving. Try it. Maybe you'll get hooked.

8
Hormonal Changes
The Impact on Women's Brain Health

HORMONES HAVE A powerful influence on how our brains develop as well as on our feelings and behaviors and personality traits throughout life. After adolescence, hormone levels remain fairly steady in men, but in women, they fluctuate like clockwork. The constant shifting affects every woman differently, and this must be factored in to any discussion of brain health. Exercise is particularly important for women because it tones down the negative consequences of hormonal changes that some experience, and for others, it enhances the positive. Overall, exercise balances the system, on a monthly basis as well as during each stage of life, including pregnancy and menopause.

The average woman has four hundred to five hundred menstrual cycles in her lifetime, each one lasting four to seven days. If you add them all up, it comes out to more than nine years — a long time for women who suffer premenstrual syndrome (PMS). "You can't be bitchy and agitated and short-tempered and have a decent life," says a thirty-eight-year-old colleague I'll call Patty. "I know the feminists hate it when you say this, but some of us do go crazy."

Crazy isn't the word I would use, but it does capture the frustration many women feel when their hormones take over. Approximately 75 percent of women experience some form of premenstrual

191

distress, physical or emotional or both, and Patty is among the subset for whom the symptoms can be severe enough to disrupt their lives (14 percent miss school or work at some point because of PMS). Every month since she was about sixteen, if she doesn't exercise Patty gets tired, irritable, itchy, anxious, agitated, and aggressive in the days leading up to her period. She has difficulty focusing; she tosses and turns at night; and she craves carbohydrates. Her ankles and belly swell; her face develops a rash; she gets constipated; and her breasts hurt. "That's when I really push myself," she says. "The week before my period I *have* to do an hour of cardio four days a week or I can't stand myself."

She learned early on that aerobic exercise dramatically dials back her symptoms. Just shy of five foot ten with a shock of red hair and a bright, broad smile, Patty worked for the Elite modeling agency from childhood through her early twenties. She wasn't into sports, but beginning in her teens, she exercised almost manically, sometimes three hours a day, to keep her weight down to 110 pounds. Without the exercise, her mom found her impossible to deal with. The absurdity of sticking to that weight convinced her to quit modeling, and she's since earned a master's degree in social work. There have been times, even years, when she's lapsed from her routine, but she always comes back to it. "It helps the most with the mood swings," she says. "It takes the edge off and gets out that aggression that comes with the hormones."

Normally tolerant and easygoing, Patty says she gets "snappy" and has a short fuse during PMS. "Her radar goes on hypersensitive," says her husband, Amon, an architect with neat black hair and rimless eyeglasses. "Her sense of smell, sound, light. Her sense of order. She becomes extremely particular. For example, she wants me to be around her, but I must be around her in a very specific way."

"He'll be next to me on the couch," she says. "I'll hear him breathing and be like, Do you have a sinus infection or something?"

"Exactly!" he says, laughing. "And she might ask if my father had sinus problems, and then it becomes an entire discussion about my family's nasal history."

Patty and Amon are conscientious people who generally communicate well and support each other, which is extremely important for women going through hormonal changes. Amon suggests that they go to the gym together on days she'd rather skip it. "Patty is one of those people in whom you can see the storm clouds building before it comes," he says.

The term PMS became politicized in the 1970s because some felt that it labeled a natural aspect of women's lives as a medical issue and created a perception that all women have a psychiatric disorder once a month. The subject was hotly debated by the medical experts who decide what should be included in the *Diagnostic and Statistical Manual* and what each condition should be called. PMS has been renamed in various editions of the *DSM*, and in 1994 the entry was changed from the inscrutable late luteal phase dysphoric disorder (LLPDD) to premenstrual dysphoric disorder (PMDD). But the requirements for a medical diagnosis of PMDD are so stringent that they leave out the majority of women who cope with what we all think of as PMS. Call it what you like, but to me the issue is whether any of the 150 symptoms in the *DSM* interfere with your quality of life.

PMS: NATURAL UPS AND DOWNS

Scientists don't know precisely what causes PMS, but changes in hormone levels are an obvious place to look for clues. The sex hormones are powerful messengers that travel throughout the bloodstream, and aside from overseeing the development of gender characteristics, they influence the brain in many ways. The cycle begins with signals from the hypothalamus that prompt the pituitary gland to secrete hormones called gonadotropins, which travel to the ovaries and trigger the mass production of estrogen and progesterone.

Estrogen peaks at five times its baseline level just before ovulation and then follows an up-and-down pattern for the two weeks leading up to the period, after which it evens out. Progesterone begins a fitful rise after ovulation (to about ten times its lowest level) and peaks just before menstruation. During pregnancy, estrogen skyrockets to fifty times normal levels, and progesterone increases tenfold. Then during menopause, both hormones decline until they nearly disappear.

The difference between women who suffer from PMS or postpartum depression or tumultuous menopause and those who don't seems not to be a matter of the level of these hormones but, rather, of the body's sensitivity to other neurochemical changes they trigger.

For instance, in relation to mood, as well as overall brain function, hormones play an important role in regulating neurotransmitters. Both estrogen and progesterone create more receptors for serotonin and dopamine throughout the limbic system and thus increase these neurotransmitters' effectiveness. And in just the past few years, scientists have discovered that estrogen signals the production of brain-derived neurotrophic factor (BDNF), which in turn creates more serotonin. While there is much we still don't know about the complex interplay between shifting hormonal levels and brain function, this link to the neurotransmitter system is shaping up to be pivotal.

In one 2004 study researchers used PET scans to compare neurotransmitter activity in women with and without PMDD. They found that the brains of women with the diagnosis had an impaired ability to "trap" tryptophan in the prefrontal cortex, thus limiting the production of serotonin, which helps regulate mood and behavior such as angry outbursts. In another study, psychiatrists from London's Kings College purposely depleted tryptophan in a group of women during their premenstrual phase and found that it led to more aggressive behavior when provoked. These were healthy women with no PMS symptoms or mood issues at the time. Each woman was told that if she reacted to a computer cue faster than a competitor in another room, she could adjust the volume of an annoying sound that would penalize the other woman. If she lost, however, she would get buzzed.

In fact, there was no opponent. All of the participants were subjected to the noise—which kept getting louder—half of the time. As the volume increased, the women with depleted tryptophan aggressively cranked up the volume, lashing out at their imaginary opponent. The study concluded that lowering the precursor to serotonin in healthy women increases their tendency toward aggressiveness. "They were much more likely to retaliate than women with normal levels of serotonin," says Alyson Bond, who conducted the study. "They behaved in a similar way to habitually aggressive people."

Aggression is just one symptom, and the story of PMS—like that of depression—is more than a one-neurotransmitter drama. A long chain of events connects the production of a hormone to the signal that manifests as feelings or behavior, and any broken or damaged link can spin the outcome in another direction. This is just one reason why PMS, pregnancy, and menopause affect every woman differently. It's impossible to say where the gap is in Patty's brain chemistry, for example, but there's no question that exercise helps close it. "It's almost like I'm in a fog before my period," she says. "I could take my ADHD medication, and it won't do a damn thing. Exercise helps clear my head."

RESTORING BALANCE

Exercise isn't necessarily the only answer if you suffer from PMS, but it can dramatically reduce the symptoms and give you a handle on a part of life that feels beyond your control. And with a lifestyle change, medication may not be necessary.

Many women already know this: one survey of more than eighteen hundred women found that at least half of them use exercise to alleviate the symptoms of PMS. In addition to reporting less physical pain, the women who exercise scored better on evaluations of concentration, mood, and erratic behavior.

The notion that exercise alleviates physical symptoms both before and during menstruation is far more accepted than its proposed effects on mood and anxiety. Frankly, there is little experimental evidence to prove specifically that exercise helps the mental symptoms of PMS. Perhaps the best study on the subject comes from the Duke University lab of James Blumenthal, who pioneered much of the research on exercise and depression, and it dates back to 1992.

With a small group of middle-aged women (premenopause), Blumenthal compared how aerobic exercise and strength training affected PMS symptoms. Each group trained for an hour three times a week. The twelve aerobic women did thirty minutes of running at 70 to 85 percent of their aerobic capacity, along with a fifteen minute warm-up and cool-down. The other eleven used weight machines for supervised strength training. Both groups' physical symptoms improved, but the runners improved significantly more on the mental side. They felt better on eighteen of twenty-three measures, the most significant of which were depression, irritability, and concentration. The sharpest distinction was that the aerobic set showed a less pessimistic outlook and more interest in the world.

One explanation, certainly, is that physical activity increases levels of tryptophan in the bloodstream and thus concentrations of serotonin in the brain. It also balances dopamine, norepinephrine, and synaptic mediators such as BDNF. By stabilizing such a broad number of variables, exercise helps to tone down the ripple effects of shifting hormones.

Exercise also ties into a more nuanced theory of PMS that is just developing. Estrogen and progesterone both get transformed into dozens of hormonal derivatives, some of which are of great interest to neuroscientists because they regulate the brain's major excitatory and inhibitory neurotransmitters—glutamate and gamma-aminobutyric acid (GABA). During the hormonal fluctuations of the premenstrual phase, the levels of these derivatives relative to one another can get out of whack, which can lead to too much excitement of the nerve cells in the brain's emotional circuitry. This could happen because too much glutamate is being produced or not enough GABA, but either way, the runaway activity can lead to mood changes, anxiety attacks, aggression, and even seizures.

One recent study found that while hormone levels were the same in women with and without PMS symptoms, their GABA levels were different. Exercise has widespread effects on the GABA system, which puts the brakes on excessive cellular activity as Xanax does. Studies in rats have shown that just one bout of exercise turns on the genes that produce GABA, for instance. Exercise restores the balance between the opposing forces of activity in the brain during a time that is tumultuous for some women. It also fine-tunes the hypothalamic-pituitary-adrenal (HPA) axis, which you may recall from earlier chapters improves our ability to cope with stress. And, not to be overlooked, exercise improves energy and vigor, which impacts all of the other symptoms.

PREGNANCY: TO MOVE OR NOT TO MOVE?

No myth about women's health has existed for so long as the belief that women should stop exercising during pregnancy. Perhaps because childbirth was a life-threatening event before modern medicine, pregnancy was considered a period of confinement—a time to stay home, reduce activity, and rest in bed. It might be dangerous to disturb the unborn child. Exercise? Out of the question.

It's only recently that doctors have begun to shift their thinking. In 2002 the American College of Obstetricians and Gynecologists (ACOG) began recommending at least thirty minutes a day of moderate intensity aerobic exercise for pregnant and postpartum mothers. It's a potentially powerful guideline, given that 23 percent of active women stop exercising when they become pregnant. Equally important, though, is that for the first time the ACOG recommended that sedentary women *begin* exercising when they become pregnant, largely to counter risks such as diabetes, high blood pressure, and preeclampsia that can develop during gestation and harm both mother and child.

Certainly there are complications for which bed rest is the sensible prescription, so it's important to speak with your obstetrician before beginning an exercise regimen. Forget ice hockey, racquetball, basketball, and any other contact sports. The same goes for riding horses, mountain biking, practicing a balance beam routine, or doing anything where falling is part of the game. And scuba diving too. Keep in mind, however, that doctors tend to be conservative. In its 2002 recommendations, the ACOG warns against exercise for pregnant women who are overweight, diabetic, heavy smokers, or who have high blood pressure — the very women who need exercise. In these cases, exercise may not be totally out of the question; it's just that they should start very slowly, working closely with their doctors.

Many expectant mothers don't have a clear idea of what they can do, and they think in terms of avoiding rather than engaging. If they understood the benefits of exercise — not only in reducing pregnancy risks but also in improving their physical and mental health and that of their babies — then they'd feel much more comfortable being active. The truth is we don't have all the answers about the effects of exercise on pregnancy, but we do have some good ones.

During pregnancy, estrogen and progesterone levels remain exponentially higher than normal, and in some cases this stabilizes mood and alleviates anxiety and depression. Indeed, pregnancy can change a number of different systems for the better. For instance, some women with attention-deficit/hyperactivity disorder are surprisingly able to sit still and read when pregnant. The body's reaction to hormones is specific to the individual, though, and some women experience distress.

Whatever the body's reaction, physical activity lowers stress and anxiety and improves mood and overall psychological health during pregnancy. A 2007 study from England evaluated the effects of a single bout of exercise on the mood of sixty-six healthy pregnant women who were divided into four groups. They either walked on a treadmill, swam, took an arts and crafts class, or did nothing extra. Women in both exercise groups improved their moods, even though they weren't necessarily problematic to begin with.

It's also well established that an expectant mother's state of mind may alter her baby's development. Stress, anxiety, and depression can have a frighteningly powerful impact on a pregnancy, and, in the extreme, can result in miscarriage, low birth weight, birth defects, or death of the baby. Babies born to unhappy mothers are fussier, less responsive, harder to soothe, and have unpredictable sleep patterns. And in follow-up tests, these babies are more likely to be hyperactive and suffer cognitive impairments. In rodent models, pups born to mothers subjected to stress during pregnancy (by way of shocking their feet) are skittish, clumsy, and less adventurous. And their stress regulation systems are forever altered, leaving them more vulnerable to future problems. Psychiatrist Catherine Monk, of Columbia University, has correlated these changes in human subjects. She found that when pregnant mothers with clinical anxiety are asked to participate in a stressful event, such as making a short speech in front of a group, their fetuses' heart rates are overreactive and don't calm down as quickly as fetuses of mothers without clinical anxiety. This is a sure sign that the HPA axis isn't regulating itself correctly, which means cortisol is on the loose. Also, twitchy HPA axis is a risk factor for future psychiatric issues.

Despite the fact that exercise can prevent a lot of unnecessary complications, many women are still leery of exercising while they're pregnant: surveys suggest that up to 60 percent remain inactive.

In general, studies report that exercise reduces nausea, fatigue, joint and muscle pain, and fat accumulation. Exercise cuts in half the risk of developing abnormal glucose levels, which can lead to gestational diabetes—a condition that results in overweight babies and prolonged labor. High glucose is also a risk factor for obesity and type 2 diabetes in both the mother and the baby, and these physical conditions are bad for the brain. Fortunately, exercise helps regardless of how active a woman was before pregnancy. One study showed that briskly walking five hours a week reduces the risk of gestational diabetes by 75 percent.

Several years ago a group of German researchers decided to test whether exercise would have any impact on the painful process of labor. They brought a stationary bicycle into the labor suite. Somehow they found fifty women who agreed to pedal for periods of twenty minutes, rate their pain levels, and have their blood tested for endorphins right up until they gave birth. Most of them (84 percent) said contractions were less painful during exercise than at rest, and their ratings were inversely proportional to endorphin levels. The researchers concluded that, "exercising on a bicycle ergometer during labor seems to be safe for the fetus, a stimulus to uterine contractions, and a source of analgesia."

DON'T FORGET THE BABY

James Clapp, an obstetrician and professor of reproductive biology at Case Western Reserve University, has been studying how exercise affects the child for more than twenty years. His 2002 book,

Exercising through Your Pregnancy, is a largely positive endorsement built on long-term studies he's conducted with several hundred women. He begins by dispelling the myth that exercise is dangerous, noting that in his research there were no differences in weight or skull size between babies born to active and sedentary mothers. With exercise, the fuel line between mother and baby grows, to ensure the fetus gets the nutrients and oxygen it needs. Studies from Clapp and others have shown that newborns of active women are leaner, which you might think is cause for concern, except that the physical differences even out within the first year.

Exercise seems to be more than just not harmful, though. In one study, Clapp compared thirty-four newborns of exercisers to thirty-one of sedentary mothers five days after birth. There's only so much you can do to gauge behavior at this early stage, but the babies from the exercise group "performed" better on two of six tests: they were more responsive to stimuli and better able to quiet themselves following a disturbance of sound or light. Clapp sees this as significant because it suggests that infants of exercising mothers are more neurologically developed than their counterparts from sedentary mothers. He theorizes that physical activity jostles the baby in the womb, providing stimulation not unlike the effects of touching and holding newborns, which clearly improves brain development. In another comparison at five years of age, he found no differences in behavior and most cognitive measures, but there were statistically significant differences in IQ and oral language skills. The children of exercisers performed better, and his unpublished observations suggest that years later their academic performance is better than kids whose moms were inactive, which is amazing.

There's no way to conclude just yet why this is the case in humans, but there are tempting clues from lab rats. Most intriguing is a 2003 study showing that rat pups born of exercising moms had higher levels of BDNF immediately after birth, and also at fourteen and twenty-eight days. At the same time, they performed better than controls on learning tasks related to the hippocampus. Essentially, they learned better and faster than rats born of sedentary mothers. One study showed that, for some reason, rat pups of running mothers had fewer neurons in the hippocampus at birth, but they bounced back and outpaced their counterparts. By the end of the first six weeks, the exercise group had 40 percent more cells in the hippocampus. A study in 2006 found that forcing pregnant rats to swim ten minutes a day resulted in more BDNF, greater neurogenesis, and improved short-term memory in their pups. In short, when pregnant rats exercise, neurons in their fetuses' brains are better able to link up to one another.

Although these findings can't be applied directly to humans, they certainly fit within the framework of what we've learned about exercise and the brain in the past decade. We can't say that running when you're pregnant will get your daughter into the best colleges, but on the other hand, these findings suggest that staying physically active improves neurotrophic support for the baby's brain cells. And, as you'll recall from earlier chapters, such changes optimize learning, memory, and overall state of mind. For me, the idea that exercising while you are pregnant might have an impact on the *future* of your baby's brain — that's powerful.

In another fascinating line of inquiry, researchers have studied the effects of exercise in combating fetal alcohol syndrome, a devastating disorder that results in stunted growth, retardation, and facial disfigurement. It's the leading preventable cause of birth defects in the United States, and some studies have shown that if a pregnant woman drinks even moderately, learning, behavioral, and social problems can follow for the baby. The brains of rats born to ethanol-fed mothers have lower levels of BDNF, neurogenesis, and neuroplasticity. The hippocampus is atrophied, and consequently, the pups can't learn or remember very well. Beyond the hippocampus, alcohol damages glutamate synapses, and this has wide-ranging effects in the brain.

In 2006 researchers from neuroscientist Brian Christie's lab at the University of British Columbia examined the neurological effects on rats of prenatal exposure to ethanol, and then tested the effects of exercise on those changes. As expected, the pups whose moms consumed ethanol had markedly lower rates of neurogenesis and neuroplasticity. After the pups were born and able to exercise, though, the activity reversed the brain damage to normal, which was startling.

The findings have already had an impact on how doctors recommend handling babies with fetal alcohol syndrome. Parents used to be advised to keep the environment quiet and dark so as not to overstimulate the babies. Now it seems that it's better to provide physical stimulation and activity, to give the babies' brains a chance to counteract their neurological deficits.

It never ceases to amaze me how our brains can repair themselves if we keep our bodies moving, as they were designed to do.

POSTPARTUM DEPRESSION: A BOLT OF THE BLUES

Tony and Stacy were desperate. It was a rainy Friday afternoon when the couple decided they needed an elliptical trainer *right now,* but the NordicTrack store at the mall was out of stock. They had to locate a seedy Boston warehouse to pick it up, and then the seats of their SUV wouldn't lay flat, so the trainer stuck out the back as rain poured down during the drive home. It must have weighed two hundred pounds, that soggy box, and wrestling it inside the house was just the beginning of Tony's work for the evening.

"We got home, and I immediately had to put it together," he says. "It isn't my strong suit, but at that point I just wanted her to be better."

They needed a treatment for Stacy's postpartum depression, which had come out of nowhere after she gave birth to their first child, a son named Carter. For five months now she had been exhausted but unable to sleep soundly; she felt guilty if she left her infant alone; she hated her body; she lost interest in the world; and she was prone to bursting into tears without warning. These symptoms are unrelated to the temporary blues most women experience within the first weeks after delivery, and they're more common than most people realize. For 10 to 15 percent of new mothers like Stacy, everything seems fine at first, but then postpartum depression strikes, and it can stick around for a year or more. When I mention to my medical and psychiatric colleagues that such a significant number of new mothers suffer postpartum depression—a fact I only learned in researching this book—they are as shocked as I was.

Antidepressants are the typical treatment, but the Lexapro Stacy had tried made her feel numb to everything. She only stayed on it for a few days and was wary of trying another medication. As she and Tony sat in my office on that rainy afternoon, I explained how aerobic exercise works as well as or better than drugs for some people with depression. I only wish all of my patients were as responsive: when they left, they made a beeline to the mall to buy the trainer.

Tony finished putting it together late that evening, and Stacy jumped right on it for a twenty-minute workout.

"Those things are hard when you first start!" she recalls. "I knew it was doing something—you get that burn."

"I think that's what grabbed her at first," Tony says. "The burn and the feeling that it was going to help her appearance. I don't think she grasped right away that it was helping her mentally and helping her sleep."

"No, I didn't."

"I noticed. I said, Stacy, it's like night and day. It really is. The first thing it changed was her quality of sleep—"

"Which of course made me feel better during the day."

"And then her mood followed."

"I had a *lot* more energy. I would feel better when I got off than when I got on it. Even now, after playing with Carter all day long, I'll feel exhausted, but I make myself do it anyway. And I'm in a better mood and am happier and feel more energetic."

What makes Stacy's story dramatic is that before she gave birth, at age twenty-nine, she could most aptly be described as bubbly. She had no history of depression and was "the happiest person" Tony had ever met. They are an incredibly loving couple, young yet old-fashioned, and, above all, they've always had a lot of fun together. But after Stacy's pregnancy, she says, "everything switched around."

She is tall, blond, athletically trim, and was only five pounds over her normal weight two weeks after giving birth. She looked terrific, which is relevant here only because she didn't see it that way. On the rare occasions when they went out after Carter was born, Stacy would try on a dozen outfits. "I thought I looked horrendous," she says. "No matter what anybody said, I didn't believe it in my heart of hearts."

"No fooling around; it was nine or ten different shoes and shirts and pants," Tony says. "She saw a different person in the mirror."

But there was more at play than a negative self-image. After the initial excitement of bringing Carter home, fatigue set in, and with it came a raft of crummy feelings. Stacy stopped expressing opinions or interest in anything. She moved the crib into their bedroom and woke up every few hours to check on the baby. "I never wanted to leave Carter," she says. "And I'd always feel guilty if I did."

New mothers overcome by depression start to question themselves and wonder if there's something wrong with them. If they're having trouble, they assume they must be terrible mothers. The instinct is to push away from the world, the baby in particular, and this leads to inner conflict and self-flagellation. Here's this biological purpose for your being and you're ashamed that everything is not blissful, convinced you're the only mother in the world having such feelings. This event that is supposed to be so wonderfully life-fulfilling instead triggers a black cloud.

It took several months of Tony gingerly raising the subject before Stacy recognized that something was off. "I didn't feel like me anymore," she says. "I had no idea how I was going to get back to that."

Stacy had lifted weights occasionally, but as I explained to her and Tony, aerobic exercise is different—and crucial for mood. Now she spends forty-five minutes on her elliptical trainer almost every night. If she misses it for more than a few days, she has trouble sleeping and notices a drop in energy and mood. Does that mean she's still depressed and just masking it with exercise? Not exactly. It's just that if symptoms flare up, as they sometimes do during her period, she hits the trainer to make sure they don't snowball into something worse. Above all, she knows she can handle it. "If I exercise, I'm fine," she says. "I feel like I'm back to normal."

GETTING BACK OUT THERE

Scientists know plenty about how aerobic activity curtails the symptoms of general depression (see chapter 5), but new mothers warrant special consideration. It's not so much the increase in hormones that causes postpartum depression, research suggests, but the effects of withdrawal when they plummet after childbirth. In 2000 Miki Bloch of the National Institute of Mental Health published a study in the *American Journal of Psychiatry* in which her lab re-created the hormonal conditions of pregnancy in two groups of thirty-something mothers: one with a past history of postpartum depression and one without. (Neither group of eight women had symptoms of depression during the study.) All of the women were given pills to stimulate the production of estrogen and progesterone, and after eight weeks the hormones were secretly replaced with a placebo. The effect was dramatic. During estrogen withdrawal, five of the eight women with a history of postpartum depression experienced a relapse of symptoms; the other group didn't notice a thing.

Given how powerfully hormones affect neurotransmitters, Bloch proposed that some women's brains simply can't compensate for the sudden changes or that the normal signals are amplified in a way that disrupts mood. From this perspective, exercise might be even more effective for new moms experiencing depression than for the general population because it normalizes neurotransmitter levels.

The best study on this issue was conducted in Australia several years ago, with twenty women suffering postpartum depression who'd given birth within the previous year. Half of them were on antidepressants. Researchers chose a form of exercise that is exceedingly convenient for new moms: walking with a stroller. One group of ten women walked with their strollers for forty minutes at 60 to 75 percent of their maximum heart rate three times a week and attended one social support meeting, while the other ten women, in the control group, carried on their normal routines. They all established a baseline score on the Edinburgh Postnatal Depression Scale (EPDS) and were tested again at six weeks, and then at twelve weeks, when the trial ended. Anyone who scores higher than 12 is considered clinically depressed. The stroller-pushers increased fitness and significantly lowered their EPDS scores at both intervals. The exercise group started with a mean score of 17.4, and it dropped to 7.2 and then 4.6. The control group started with a mean of 18.4, dropped to 13.5, and then nudged up again to 14.8.

Statistically, fit mothers have a lower incidence of depression. In one survey of one thousand women in the South of England six weeks after childbirth, the 35 percent who reported doing vigorous exercise three times a week had significantly fewer mood problems. They also had lost more weight, stayed more socially active, and felt more confident and satisfied in being mothers. An exercise routine can help new moms reestablish control over their lives and keep them from feeling overwhelmed. It also provides a great way for them to take time for themselves, which is important in staving off resentment. Like Stacy, about 70 percent of women are dissatisfied with their bodies at six months after childbirth, and obviously exercise can get them back in shape and boost self-image.

Unfortunately, the message that exercise provides something more than physical redemption has been slow to reach doctors and their patients. "People think of exercise in terms of physical health, but not mental health," says Jennifer Shaw, an obstetrician-gynecologist in Brookline, Massachusetts, who is a clinical instructor at Harvard Medical School. "It's difficult as a physician to get people to take exercise seriously, as a treatment that actually has medical benefits aside from taking off the pounds."

The field of obstetrics isn't really set up to diagnose or treat mental health issues related to pregnancy. Shaw will bring up exercise as a solution to a problem, but she says it's hard for any physician to find the time to discuss preventive medicine. It can also be dicey, she points out, to suggest exercise to a woman who's juggling so many new responsibilities and might not be feeling so hot about her body. "The first thing to drop off the list when life gets more complicated for women is exercise," Shaw says. "I don't think there's an appreciation for what it does, but I do think it has a stabilizing effect on mood."

The worst advice for new mothers who are feeling down is to take it easy. Rest is important, certainly, but not as important as activity. New mothers need support from their husbands to carve out the time to work on their bodies—and their brains—as soon as possible.

MENOPAUSE: THE BIG CHANGE

Strictly speaking, menopause is a one-day event that marks the end of the twelfth month after a woman's final period. More practically, it represents the span of hormonal changes surrounding that moment. As the ovaries become less reliable with age, production of estrogen and progesterone becomes sporadic and falters. When these hormones fall out of phase, the brain's delicate balance of neurochemicals gets disrupted.

Symptoms typically kick in several years before menopause, between the midforties and midfifties (the median age for menopause is fifty-one), and might last several years afterward. They include the so-called vasomotor symptoms of hot flashes and night sweats as well as irritability and mood instability. And as with the other hormonal shifts I've discussed, there's no predicting how someone will be affected—some pass through menopause without really noticing, and others are tormented. Most women experience at least a few of the symptoms, and many of those who exercise find that it helps. The great value of exercise for women beyond menopause is that it helps balance the effects of diminished hormones, and as you'll see in the following chapter, it protects against cognitive decline. From an evolutionary perspective, exercise tricks the brain into trying to maintain itself for survival despite the hormonal cues that it is aging.

Exercise also provides protection, lost to the ebb of natural hormone levels, against health problems such as heart disease, breast cancer, and stroke. It's rare for premenopausal women to have heart attacks unless they have a genetic predisposition or complications such as obesity or diabetes. This has always been the rationale behind hormone replacement therapy (HRT): estrogen and progesterone protect women against chronic disease, so these hormones need to be replaced after menopause. In recent years, however, this assumption has been overturned, and now many doctors simply won't prescribe HRT.

The controversy erupted in 2002 when researchers at the National Institutes of Health noticed some alarming statistics for a group of postmenopausal women participating in one of a series of

Women's Health Initiative studies. The women undergoing HRT had a 26 percent higher risk of breast cancer, a 41 percent higher risk of stroke, and a 29 percent higher risk of heart attack.

In the wake of this disturbing news, millions of women stopped taking hormones, and the *New England Journal of Medicine* published a population survey showing that in 2004 breast cancer rates dropped 9 percent. Then a prominent British study reported that women on HRT have twice the risk of developing dementia, which is a major concern for anyone beyond middle age. However, there are studies supporting the use of HRT for short periods during menopause. The only universal advice for menopausal women is to ask your doctor. Whatever the answer, the contradictions put many women in a painful bind.

CONSIDER THE SYMPTONS

The most common reason women seek HRT is to alleviate the physical symptoms of menopause, specifically hot flashes, and nobody disagrees that it works wonders on this front. Exercise is one alternative, although evidence for its effects on hot flashes and night sweats is inconclusive. Several large observational studies, one of which included sixty-six thousand menopausal Italian women, have shown that lower levels of exercise correlate to more vasomotor symptoms, but other studies have shown no association.

Some ob-gyns will tell you that exercise actually triggers hot flashes. At least with exercise you can safely conduct your own experiment without fear of long-term side effects. Either it helps relieve your symptoms or it doesn't, but you don't have to worry about it undermining your health down the line. What gets lost in the question of whether exercise helps menopausal women with hot flashes is the big picture, namely that it guards against heart disease, diabetes, breast cancer, and cognitive decline.

The physical symptoms of menopause exacerbate the mood symptoms, and there's no question exercise helps in this regard. One woman told me that the most frustrating part of aging is that she feels like her body is out of control. She gained weight, suffered hot flashes and high blood pressure, and her vision deteriorated. On top of that, she feels anxious and depressed at times. What exercise provides is a sense of control, over the physical changes but more so over the emotional ones. "I know that exercise helps to contain many of these symptoms," she says. "It helps guide me, so that I can be as proactive as possible in dealing with some of these things that are so out of my control."

As with PMS, in menopause it seems to be the fluctuations, not the levels, of hormones that leave some women vulnerable to anxiety and depression. Women are twice as likely as men to suffer from anxiety and depression to begin with, and that risk increases further when they enter menopause, according to a study conducted by psychiatrist Lee Cohen, a women's health specialist at Massachusetts General Hospital. As part of a larger effort called the Harvard Study of Moods and Cycles, he followed 460 women, thirty-six to forty-five years old, for six years to compare mood changes as they entered menopause. None had any history of depression, but their risk of developing it doubled during menopause.

In a recent survey of 883 women (ages forty-five to sixty), researchers from the University of Queensland in Australia found a strong correlation between exercise and menopausal symptoms. An astounding 84 percent of the women reported that they exercised two or more times per week, and they had significantly lower rates of the physical and mental symptoms of depression than nonexercisers. Specifically, they felt less tense, tired, and fatigued. They reported fewer headaches and less tightness or pressure in their bodies. Overall, the study concluded, exercise can have a tremendous impact on a woman's sense of well-being and quality of life.

EXERCISE REPLACEMENT THERAPY

It's well established that more women suffer from Alzheimer's disease than men, even when the statistics are adjusted for the fact that women live longer. On the other hand, the protective effects of exercise on cognitive decline seem to be magnified among women. In a 2001 study published in the *Archives of Neurology*, Danielle Laurin of Quebec's Laval University analyzed the relationship between exercise and physical activity among a group of 4,615 elderly men and women over the course of five years. Laurin found that women over sixty-five who reported higher levels of physical activity were 50 percent less likely than their inactive peers — women and men alike — to develop any form of dementia.

Until the Women's Health Initiative came along, scientists believed that HRT protected against cognitive decline, but the evidence doesn't support that conclusion. Now, one of the questions researchers have begun to tackle is whether exercise and hormones have interactive effects on cognitive decline after menopause. Research from Carl Cotman's lab at the University of California, Irvine, suggests that estrogen is necessary for exercise to increase levels of BDNF in the prefrontal cortices of female rats. But the design of the study doesn't necessarily translate to the conditions of menopause in humans — the rats' ovaries were removed at three months of age, which equates with young, healthy women. The first human reports on this question suggest that estrogen is not an essential ingredient for exercise to protect against cognitive decline. In one study, physiologist Jennifer Etnier, who is now at the University of North Carolina at Greensboro, administered tests of mental processing speed and executive function to 101 postmenopausal women and compared the results to their reported levels of regular aerobic activity. Those who were more physically active had higher scores regardless of whether they had been undergoing HRT.

The most telling study on this subject comes from the lab of psychologist Arthur Kramer at the University of Illinois, Urbana-Champaign, which has been at the forefront of correlating certain cognitive abilities with changes in brain structure identified in MRI scans. He wanted to see whether exercise and HRT interacted in their impact on executive function and the volume of the prefrontal cortex. In a complicated design, he recruited fifty-four postmenopausal women, each of whom agreed to an MRI scan, mental tests of executive function, and a treadmill test of their maximal rate of oxygen consumption (VO2 max) to measure fitness. The data was sorted into four categories based on duration of HRT. The first group had never had hormone treatment, and the rest fell into short-term (ten years or less), midrange (eleven to fifteen years), and long-term (sixteen years or more) treatment.

The results, published in 2005, showed that women in the short-term HRT group performed better on the tests and had greater brain volume than women who had never had therapy or those who had had more than ten years. Which suggests that HRT can, in fact, be protective in the short run. When aerobic fitness was factored in, it had a significant impact on measures of performance and brain volume. Better fitness seems to offset the decline women would otherwise experience if they had never had HRT or if they had had it for more than ten years.

One of the theories from rodent research is that with long-term HRT, estrogen receptors in the brain begin to break down in the hypothalamus, the area that activates the immune response. And if the hypothalamus is not working correctly, women would be more vulnerable to diseases such as cancer. Equally important, long-term estrogen treatment in rodents also causes cellular inflammation, which is a risk factor for Alzheimer's and is associated with memory impairment.

What Kramer suggests is that exercise seems to enhance the positive effects of short-term HRT, and this fits in with the neuroprotective mechanisms I've described throughout the book. Exercise sparks production of neurotransmitters and neurotrophins, creates more receptors for them in key areas of the brain, and turns on genes that keep the positive cycle spinning. That momentum is crucial for any woman, but especially from menopause onward. After all, most women live for decades without their hormones.

ESTABLISH A ROUTINE

At least four days a week, I suggest getting out there and walking briskly or jogging or playing tennis or engaging in some form of activity that will get your pulse up to 60 to 65 percent of your maximum heart rate. You want to keep it there for an hour. People always want to know what type of aerobic activity is best, and the answer is whatever is going to allow you to build it into your lifestyle. The important thing is to stick with it, and make sure you're elevating your heart rate enough to get the benefits. It's also important to mix in strength training a couple of days a week, to shore up your bones against osteoporosis.

For younger women with PMS, I would suggest five days a week of aerobic exercise at the same level, but it might be a good idea to mix in more intense bursts like sprinting on two of those days, though not back-to-back. Some of the studies suggest that higher intensity effort has a more dramatic effect on symptoms such as irritability, anxiety, depression, and mood instability. And if your symptoms are particularly bad, and you're not completely sidelined with cramps, it's probably a good idea to do something every day during the premenstrual phase of your cycle.

The advice that surprises people the most, I think, is that it's important to keep up exercise during pregnancy, a recommendation that has finally been endorsed by the American College of Obstetricians and Gynecologists. Its guidelines specify thirty minutes of moderate-intensity aerobic activity every day during pregnancy for healthy women. Obviously, it's important to get clearance directly from your obstetrician, but it's safe for most women. Likewise, I can't stress enough the importance of picking up your routine as soon as possible after the baby is born, ideally within a few weeks. Although it seems contradictory, moving will actually reduce fatigue. And for women like my patient Stacy, it melts away anxiety and depression.

When women are younger, one of the big motivations to exercise is to stay trim, and that's fine. Use whatever gets you going. But the message I want to leave you with is that even as your body changes, exercise will keep your mind firm and taught. And in this state of mental fitness, you'll be well equipped to handle the hormonal fluctuations that every woman experiences throughout her life. Not to mention the fluctuations of life itself.

9
Aging
The Wise Way

MY MOTHER WAS known as the fast walker. At five foot six, she ruled the sidewalks in our western Pennsylvania town, and people would always ask my brother and sister and me where she was running off to. She would stride to church for the early Mass every morning except Sunday, when my father drove all of us in our good clothes. It was a mile and a half each way, which was a pretty good workout at her pace, but she didn't walk to stay in shape. She walked because she liked to walk (and to compare prices at grocery stores a mile apart).

Judging from what scientists have learned about the nourishing effects of exercise on the brain, I'm certain that my mother's level of physical activity is what kept her so sharp for so long. Well into her eighties, Vern Ratey lived a full, vibrant life. Part of that was just her personality—she always had to be doing something. I remember that we bought a couch once, after she stewed for weeks over the color and size, measuring and remeasuring. The day it was delivered, I arrived home from school to find her busy sawing off the upholstered arms so that our new piece of furniture would fit as she'd wanted it to.

It was with the same unvarnished zeal that she attended to everything, whether planting tomatoes in a craggy patch alongside

217

the house or shoveling snow. She was a professional volunteer—my father wouldn't let her work—and our basement was full of donated clothing for church rummage sales, which meant we got first pick. As a second-generation working-class American from Czechoslovakia, she was certainly a product of the Depression: thrifty and stern, tough but loving.

My father, Stephen, was four years older, and he died when my mother was fifty-nine. Although it took her several years to recover, she was a resilient woman with lots of friends, and eventually she met another man and remarried in her midsixties. The two of them spent winters in Vero Beach, Florida, where he taught her how to play golf, and she learned how to swim. In the summer she'd wake up and put on her bathing suit beneath her clothes, so she could stop by the pool; her only stroke was the dog paddle, and she'd motor around for an hour at a time in the deep end. And she kept walking: to church, to the grocery store, to go dancing or bowling, and to play bridge at the senior center three times a week.

Aside from osteoporosis, her health was strong. And her wits were sharp. Whenever I called, we would have detailed conversations about her earning master points in bridge or how she should manage her money. Her second husband died when she was in her early seventies, but she kept moving.

When my mom was eighty-six, she tripped and broke her hip, which is just the sort of accident that sends about 1.8 million seniors to the emergency room every year. Although heart disease, cancer, stroke, and diabetes are the leading causes of death for Americans over age sixty-five, many of them live in fear of falling and breaking their brittle bones. Hip fractures are particularly devastating because they require months to rehabilitate, and losing mobility in such a pivotal, weight-bearing joint can dramatically reduce a person's activity level. About 20 percent of older adults who break a hip die within a year.

As for my mom, after about six months she got back on her

feet with the help of a walker, and we were able to avoid putting her in a nursing home by having a live-in aide. But it slowed her down—she was shuffling instead of walking—and her osteoporosis was progressing more rapidly, curling her spine and forcing her into a stoop. When her body slowed down, her mind followed suit: she stopped playing bridge and started watching soap operas. A friend took her to church on Sundays, but otherwise she didn't get out much. She was slipping mentally, but didn't have dementia yet—she knew perfectly well who I was, but she had less and less to say during our conversations.

Then, the following year, she fell and broke her other hip. It was crushing for me to see her immobilized, and that's when she really stopped being herself. She lost her grip on the boundaries between what was real and what was not. The soap opera characters became part of her life, and she talked to them as if they were right there in the room. She died of natural causes at eighty-eight.

KEEPING IT ALL TOGETHER

I've talked a lot about the biological connections between the body and the brain throughout this book, and nowhere are they more important than in the discussion of aging. After all, a sound mind won't do you much good if your body fails.

In 1900, the average American could expect to live to age forty-seven. Today, life expectancy is over seventy-six, and when older folks die, the reason is more likely to be chronic disease than acute illness. But those who outlive these odds face other daunting statistics: The average seventy-five-year-old suffers from three chronic medical conditions and takes five prescription medicines, according to the Centers for Disease Control (CDC). Among those over sixty-five, most suffer from hypertension; more than two-thirds are overweight; and nearly 20 percent have diabetes (which triples the chance of developing heart disease). The leading killers are heart disease, cancer, and stroke; together they account for 61 percent of all deaths in this age group.

We already know that smoking, inactivity, and eating poorly are root causes of these bodily diseases. Likewise, the latest research is clear about how lifestyle influences the mental hazards that come with aging. The same things that kill the body kill the brain, which neuroscientist Mark Mattson, of the National Institute on Aging, sees as a positive. "I think the good news — if we take it seriously — is that many of the same factors that can reduce our risk for cardiovascular disease and diabetes also reduce the risk for age-related neurodegenerative disorders," he says. The measures we would take to guard against diabetes, for example, also balance insulin levels in the brain and shore up neurons against metabolic stress. Running to lower our blood pressure and strengthen our heart also keeps the capillaries in the brain from collapsing or corroding and causing a stroke. Lifting weights to prevent osteoporosis from devouring our bones releases growth factors that make dendrites bloom. Conversely, taking omega-3 fatty acids for mental acuity strengthens our bones.

The mental and physical diseases we face in old age are tied together through the cardiovascular system and metabolic system. A failure of these underlying connections explains why people who are obese are twice as likely to suffer from dementia, and why those with heart disease are at far greater risk of developing Alzheimer's, the most common form of dementia. Statistically, having diabetes gives you a 65 percent higher risk of developing dementia, and high cholesterol increases the risk 43 percent. We've had the medical proof that exercise protects against these diseases for decades, yet, according to the CDC, about a third of the population over sixty-five reports that they engage in no leisure-time activity. My hope is that if you understand how exercise can also protect your mind, you'll take it to heart.

Some of the most persuasive evidence about the effect of exercise on the aging brain comes from a landmark research project called the Nurses' Health Study, which began surveying the health habits of more than 122,000 nurses every two years, in the mid-1970s. In 1995 researchers began cognitive testing for some of the nurses, which allowed Harvard epidemiologist Jennifer Weuve to analyze the relationship between exercise level and cognitive ability for 18,766 women between seventy and eighty-one years old. Weuve used the trove of data to tackle the question of whether being active on a regular basis throughout adult life translates into sharper mental function when we're older. The results, published in the *Journal of the American Medical Association,* powerfully underscored her hunch: women with the highest levels of energy expenditure had a 20 percent lower chance of being cognitively impaired on tests of memory and general intelligence. The median level of activity for this group translated to walking twelve hours a week, or running just under four hours total, compared with less than one hour of walking for the least active group (of five). But Weuve says you don't have to be a "super athlete" to get a benefit. "The really neat thing is that we started to see effects at modest levels of activity—we're talking about walking an hour and a half a week," she says. Even at this relatively low level, "you start to see a benefit that is significantly above and beyond the least active women."

HOW WE AGE

Getting older is unavoidable, but falling apart is not. Why is it that some people live to be one hundred with relatively few health problems, while others suffer from chronic diseases that rob them of normal mental and physical function? To understand how aging can take such divergent paths, it's useful to have a look at life and death at the cellular level.

As we age, the cells throughout the body gradually lose their ability to adapt to stress. Scientists have yet to figure out exactly why this happens, but it's clear that older cells have a lower threshold for combating the molecular stresses of free radicals, excessive energy demands, and overexcitability. And the genes responsible for producing proteins to clean up the damaging waste stop doing their job, which can lead to a cellular death spiral that neuroscientists call apoptosis. As the damage builds up, the immune system is activated and sends in white blood cells and other factors to mop up dead cells, which creates inflammation; if the swelling becomes chronic, it creates even more damaging proteins, and these are directly linked to Alzheimer's.

In the brain, when neurons get worn down from cellular stress, synapses erode, which eventually severs the connections. With the decrease in activity, the dendrites physically shrink back and wither. Losing a signal here or there isn't such a big deal at first, because the brain is designed to compensate by rerouting information around dead patches in the network and recruiting other areas to help with trafficking. There's a certain redundancy built into the system. Remember, we're talking about one hundred billion neurons, each of which might have up to one hundred thousand inputs. It's a very social network that thrives on making new connections and, as I've mentioned, is constantly rewiring itself and adapting—provided there's enough stimulation to spur the growth of new connections. As we get older, more real estate is required to carry out any given function. Wisdom, I think, is a reflection of how adept the brain is at compensating for this loss of efficiency.

If the synaptic decay outpaces the new construction, that's when you start to notice problems with mental or physical function, ranging from Alzheimer's to Parkinson's disease (depending on where the degeneration occurs). Fundamentally, cognitive decline and all neurodegenerative diseases stem from dysfunctional and dying neurons; it's a communication breakdown. Research on aging revolves primarily around the effort to "restore the nerve cells' ability to communicate and stay alive," Mattson points out. "If you can do that, then you can prevent their degeneration and therefore prevent the disease."

As the synaptic activity decreases and dendrites retract, the capillaries feeding the brain shrink back as well, restricting blood flow. It can work the other way around too: if capillaries shrink back because you don't get your blood pumping often enough, the dendrites follow suit. Either way, it's a killer—without the oxygen, fuel, fertilizer, and repair molecules carried by the bloodstream, cells die. Levels of nurturing neurotrophins—such as brain-derived neurotrophic factor (BDNF) and vascular endothelial growth factor (VEGF)—trail off as you age too, and production of the neurotransmitter dopamine slows down, undermining motor function as well as motivation. Meanwhile, the hippocampus is getting fewer and fewer new neurons to work with. Studies in rats indicate that neurogenesis slows down dramatically with age—not because fewer stem cells are born, but because fewer of the starting pool divide and go on to become fully functioning neurons (probably due to less VEGF). Most neuronal stem cells die anyway, but the number that are put to use drops from about 25 percent to 8 percent in rodent middle age (approximately fifty years old for us) and then dwindles to 4 percent in old age (meaning over sixty-five). That's to say nothing of the vast swaths of brain that don't benefit from neurogenesis. Starting at about age forty, we lose on average 5 percent of our overall brain volume per decade, up until about age seventy, when any number of conditions can accelerate the process.

People like my mother who stay involved and active as they age can slow down the degeneration. In one study of recent retirees, researchers found that those who exercised maintained nearly the same level of blood flow in the brain after four years, while the inactive group had a significant decrease. If your brain isn't actively growing, then it's dying. Exercise is one of the few ways to counter the process of aging because it slows down the natural decline of the stress threshold. "Paradoxically," says Mattson, "it's good that cells be periodically subjected to mild stress because it improves their ability to cope with more severe stress."

In addition, exercise sparks connections and growth among your brain's cell networks in the same ways I've described in earlier chapters: it increases blood volume, regulates fuel, and encourages neuronal activity and neurogenesis. Because the aging brain is more vulnerable to damage, anything you do to strengthen it has a more pronounced effect than it would on a young adult. That's not to say starting early isn't important—if you have a better, stronger, more connected brain going over the hill, it will surely be more resilient and resist neuronal breakdown that much longer. Exercise is preventive medicine as well as an antidote. Age happens. There's nothing you can do about the why, but you can definitely do something about the how and the when.

COGNITIVE DECLINE

It shows up in the little things first. As the connections in the brain break down, you have a harder time calling to mind people and places you've known. Everybody experiences this at some point—having something on the tip of your tongue but not being able to produce it. The prefrontal cortex, which is the search engine for your memory, can't call it up. The hippocampus provides other associations to try to jog your memory, but it's frustrating that you have to work so hard at something that used to happen subconsciously. This happens to most of us as we age, but the extent to which so-called mild cognitive impairment affects people varies dramatically.

It's not necessarily progressive, but if mild cognitive impairment continues unchecked it can become dementia. You start to lose track of the events that have shaped who you are, which is a terribly threatening feeling that eats away at your sense of self. A lot of people who find themselves in this position tend to retreat, unknowingly mirroring their dendrites. They don't venture out and make new connections for fear that they won't know how to react, and they withdraw from the world, either out of embarrassment or simply because they feel uncomfortable outside the familiarity of home. Either way, the result is that they get cut off from meaningful relationships, which are an important form of stimulation for the brain. Isolation and inactivity feed the cellular death spiral, and this shrivels the brain.

The erosion is most pronounced in the frontal lobe, which encompasses both the gray matter of the prefrontal cortex and the white matter of its axons, and the temporal lobe, which catalogs words and proper names and helps form long-term memories through its tight connection to the hippocampus. If the prefrontal cortex goes, it takes higher cognitive functions with it, and that's when the basic aspects of everyday life become trying. Ironically, the abilities we take for granted—being able to tie our shoes, unlock a door, or drive to the grocery store—rely on our highest order brain functions, such as working memory, task switching, and blocking out irrelevant information. That's why even a trained monkey has trouble properly buttoning a shirt, and it's probably why one of my patients always forgets to zip up his fly. At seventy-eight, he makes the same mistake over and over because—despite his wife's haranguing—his working memory can't hold onto the fact that he just used the bathroom.

The temporal lobe—our mental dictionary—is one of the areas that atrophies in Alzheimer's disease. A simple test for the disease is to show someone a list of words and ask what she can recall half an hour later.

As I mentioned in the first chapter, researchers at the University of Illinois have conducted a number of studies showing a strong correlation between fitness levels and better performance on tests that target these brain areas. In one study, older adults who reported having a strong history of aerobic exercise clearly had better preserved brains, according to MRI scans. But a correlation is merely interesting to lab scientists. They wanted to see if exercise caused structural changes in these areas.

A team led by neuroscientist Arthur Kramer divided fifty-nine sedentary people ranging in age from sixty to seventy-nine into two groups that would hit the gym three times a week for one hour over the course of six months. Members of the control group embarked on a stretching routine, and the others walked on treadmills, starting out at 40 percent of their maximum heart rate and ramping up to 60 to 70 percent. The only variable was fitness, and indeed, after six months, the walking group averaged a 16 percent improvement in their maximal rate of oxygen consumption (VO2 max), which is a measure of the lungs' capacity to process oxygen.

But the groundbreaking finding came from MRI scans before and after: those with improved fitness had an increase in brain volume in the frontal and temporal lobes. Scientists knew that this could happen in the hippocampus, but the suggestion that brain volume increased in the cortex was "out there," in the estimation of neuroscientist Carl Cotman, the researcher who pinpointed the link between exercise and BDNF. "I'm sure he's right," Cotman says of Kramer. "He's a very honest, accurate guy. But the findings are definitely on the outer fringe. I mean, I don't think anybody with animal studies has shown that a brain region in an older animal gets bigger from a very short period of physical activity."

It remains to be seen whether Kramer's findings can be replicated, but the idea that just six months of exercise remodels these crucial areas of the brain is incredibly heartening. In the scans, the exercisers' brains looked as if they were two to three years younger than they were. The resolution of the images doesn't show specifically what composed the growth, but given what we know from animal studies, Kramer has his suspicions. "It could be new vascular structure, new neurons, new neuronal connections," he says. "I think it's probably all of the above."

The major implication is that exercise not only keeps the brain from rotting, but it also *reverses* the cell deterioration associated with aging. More than likely, what Kramer's scans show is how exercise improves the brain's ability to compensate. "Let's say the prefrontal cortex isn't functioning quite up to par," he explains. "You might be able to recruit other areas of the cortex to do the task in a different way. One way to think about the increased volume is that it might turn back the clock in terms of how well the circuits function to do different things."

There's an awful lot our brains can do with an extra two to three years.

EMOTIONAL DECLINE

It's no wonder that some people get cranky with old age. It's a time often marked by loss—of career, relationships, possibilities, purpose, resilience, courage, and vitality. Depression can come out of nowhere, and it's an important issue for older adults because it increases the risk for dementia. The hormones estrogen, in women, and testosterone, in men, decrease with age, and this can lead to a shift in mood or a loss of vigor and interest. Also, one of the reasons depression is a risk factor for dementia is that it has corrosive effects on the hippocampus: if we're under constant stress and the hormone cortisol stays elevated, it eats away at our synapses. Since aging neurons are less resistant to the effects of stress to begin with, this is really something to guard against or, better yet, to attack proactively.

As we get older and our bodies weaken and our energy levels decline, we might be reluctant to take on challenges such as that trek in Nepal or even the local bridge tournament. But challenges are important because they boost our resilience.

I think back to my mom, who was so vibrant and involved until she broke her hip. It seemed that she became even more daring with age, and instead of turning down novel experiences she would respond with, "Why not!?" As a small example, one night while she was visiting, we were talking about eating at a fancy new Thai restaurant, and I dismissed the idea as quickly as I had mentioned it, assuming my mom wouldn't like it. She must've been about eighty at the time, and I couldn't picture her digging into a plate of exotic cuisine that looked like architecture. But she said, "Let's go! I want to try it." It makes me chuckle to this day, picturing her face when she sampled my curry dish—the spiciest food she'd ever tasted. She loved her coconut soup, though, and we laughed all the way through dinner.

Exercise is obviously a great way to challenge yourself and your brain, and all the better if it puts you in contact with other people and gets you out and about. Consider a recent study from the Rush Alzheimer's Disease Center. It showed that people who feel lonely—those who identify with statements such as "I miss having people around" and "I experience a general sense of emptiness"— are twice as likely to develop Alzheimer's. And it's clear from the studies at Duke University that exercise reduces depression and is even better than Zoloft at keeping people from relapsing.

A particularly important effect of exercise for older adults is that it rallies dopamine, which diminishes with age. This is a critical neurotransmitter in the context of aging because it's the major signal bearer of the reward and motivation systems. Apathy can become a defining characteristic for older folks, and it's particularly important to watch for when people move into retirement communities and nursing homes. Even in the best and homiest facilities, depression and a lack of motivation can set in as people feel they're just waiting to die.

I'm familiar with one retirement home that's doing something about this problem by trying to get the residents involved and interested in exercise. University Living, in Ann Arbor, Michigan, has a fitness center with aerobic and strength-training machines designed to accommodate people who aren't so agile anymore, even those who use walkers to get around. The gym is called Preservation Station. And they've hired an exercise physiologist who specializes in aging to run group classes and serve as a personal trainer for the able-bodied among their seventy residents. But a lot of what June Smedley does in her role as fitness director is knock on doors and entreat people to work out. "Boy, they get mad at me sometimes!" Smedley says. "They'll chase you out of their rooms." Most of her prospective trainees are eighty-somethings who didn't grow up with the notion that exercise is inherently healthy, and she says the motivation to participate is pretty low. "A lot of them are depressed, and that just drives their whole mentality," she says. "Their favorite thing to do is just go and sit."

One of her exercise poster boys is an eighty-year-old former engineer—I'll call him Harold—who lives there because his wife has Alzheimer's and needs daily care. He works out five days a week, doing a full regimen that includes a ten-minute warm-up, a round on the weight machine, balance drills on the physioball, and then a thirty-minute aerobic workout on the NuStep, a sort of recumbent stair climber with arm levers.

"I don't make a career out of it," says Harold, "but my primary motivation is to be able to do the things I enjoy doing." That means skiing in the winter and playing golf in the summer, when he walks eighteen holes twice a week. Six months after his eightieth birthday, Harold went skiing for a week in Utah with friends, in keeping with a fifteen-year tradition. He says the core strengthening he'd been doing with June helped his stamina and his form. He was able to ski nonstop from the top of Alta, at 10,550 feet, to the bottom—a descent of about 2,000 feet in elevation that would be difficult for any flatlander. In addition to keeping him on the slopes, his routine has helped him manage the strain of tending to his wife. "Even though there's good health care here, I still have a lot of husbandly duties," he says. "I think exercise reduces the stress. I always work up a pretty good sweat, and it gives me something to look forward to each day. And time for myself. I feel a sense of accomplishment. There's no question that it's helpful for my mental and emotional and physical well-being."

DEMENTIA

Dementia is a loss of function that dramatically undermines our ability to get through daily life. It happens when a particular area of the brain is impaired or shuts down, not unlike when a fuse gets blown in a home's circuit breaker: the kitchen appliances may work fine, but the lights in the bedroom have gone dark.

There are different types of dementia, depending on which circuit is down, and what tripped it. The most common form, by far, is Alzheimer's disease, which is marked by inflammation and the buildup of amyloid plaque that begins in the hippocampus and spreads to the frontal and temporal lobes, as well as intracellular waste called neurofibrillary tangles. According to the 2000 census, about 4.5 million Americans have Alzheimer's, and that number is expected to triple, to more than 13.2 million, over the next fifty years as baby boomers enter old age.

Stroke stems from a collapse or rupture or blockage of capillaries anywhere in the brain. If the blood flow is cut off to the temporal lobe—the brain's dictionary—you can speak but you can't get the words right. If you have a stroke in the frontal cortex, you won't be able to speak, but you can understand what people are saying to you.

The next most common form of dementia is Parkinson's disease, in which the dopamine neurons in the subtantia nigra get depleted and cut off the flow of the neurotransmitter to the basal ganglia, which is the brain's automatic transmission. The basal ganglia is necessary to smoothly shift between mental or physical tasks and start and stop motor movements; when dopamine runs dry, it's as if the transmission fluid has been drained, thus the classic tremors of Parkinson's. The disease typically comes on later in life and afflicts about 1 percent of the population over age sixty. (Early-onset cases like that of Michael J. Fox are rare.) The motor impairments show up first and are followed by mental ones, such as depression, attention problems, and, ultimately, dementia.

The biggest risk factor for dementia is the set of genes we're born with. There are a number of genes related to Alzheimer's, such as the apolipoprotein (Apo) E4 variation, but it's important to remember that having a certain gene doesn't necessarily predetermine our fate. The ApoE4 variation, for instance, is present in approximately 40 percent of Alzheimer's patients, but 30 percent of the general population (unafflicted by the disease) carries it too. And there are plenty of Alzheimer's patients who do not have the ApoE4 variation. Genes determine our risk for a disease, but our lifestyle and environment can either trigger or suppress those risks. One study, for example, showed that our chances for developing Alzheimer's drops 17 percent for every year of education we have beyond high school.

Statistics aside, we know from animal studies that exercise can redirect the biology of the brain. Carl Cotman tested the effects of exercise on mice bred with a gene predisposing them to plaque buildup and found that exercise slowed down the accumulation compared to the inactive mice. Exercise also prevents inflammation, which Cotman believes might trigger the plaque accumulation—inflammation increases in the transition stage from cognitive decline to Alzheimer's.

Mattson found the same sort of result in rats that had their dopamine neurons knocked out to mimic the biology of Parkinson's. The brains of animals allowed to run on wheels showed better plasticity and more connections in the basal ganglia, suggesting that they had adapted by building up circuits to compensate for the dopamine decline.

But our knowledge of what exercise does for Parkinson's, in particular, extends far beyond lab rats. Over the past five or ten years, exercise has been used more and more as a treatment, particularly in the early stages of the disease. Researchers started testing the effects of exercise because it calls into action the motor area that degenerates in Parkinson's; stimulating the basal ganglia through exercise increases the connections and boosts BDNF and other neuroprotective factors. One study looked at the effects of using exercise in combination with levodopa (L-dopa), the common drug treatment for Parkinson's and a dopamine precursor that increases production of the neurotransmitter. The problem with L-dopa is that it loses effectiveness over time (and it has many side effects). Doing forty minutes of easy stationary cycling immediately before taking L-dopa improved the effectiveness of the medicine on motor function.

And while researchers can't say exactly how exercise counteracts the effects of Alzheimer's—they're still trying to figure out what causes the disease—Cotman believes that reducing inflammation and boosting neurotrophic factors are likely explanations.

Population studies support the evidence that exercise holds off dementia. In one, about fifteen hundred people from Finland originally surveyed in the early 1970s were contacted again twenty-one years later, when they were between sixty-five and seventy-nine years old. Those who had exercised at least twice a week were 50 percent less likely to have dementia. What's particularly interesting is that the relationship between regular activity and the onset of dementia was even more pronounced among those carrying the ApoE4 gene. The researchers suggest that one explanation might be that their brains' neuroprotective systems are naturally compromised by the gene variant, making lifestyle particularly important. The bottom line, as Mattson says, is that "all we can do at the present time is modify the environmental factors to get the best out of whatever genes we have."

THE LIFE LIST

Much of the public discourse on aging focuses on baby boomers becoming senior citizens and the belief that their vast numbers will take an unprecedented toll on the health care system, in the form of dementia and other costly health problems. But I don't believe we're stuck with this picture of doom and gloom. Despite my generation's familiarity with fast food and pay per view, we also came of age with Kenneth Cooper's revolutionary concept of aerobics. Unlike previous generations, we recognize how a healthy heart and healthy lungs stave off disease, and we know our way around the gym. My mother just happened to have the good habit of walking, and even Harold, the eighty-year-old skier from Michigan, isn't terribly well versed in matters of health and fitness. He once asked the trainer June Smedley what was causing a muscle twitch, and when she suggested it might be dehydration, he scoffed, saying, "I drink lots of fluids—coffee, milk, and wine!"

I have faith that when people come to recognize how their lifestyle can improve their health span—living better, not simply longer—they will, at the very least, be more inclined to stay active. And when they come to accept that exercise is as important for the brain as it is for the heart, they'll commit to it. Here's how exercise keeps you going:

1. It strengthens the cardiovascular system. A strong heart and lungs reduce resting blood pressure. The result is less strain on the vessels in the body and the brain. There are a number of mechanisms at work here. First, contracting muscles during exercise releases growth factors such as VEGF and fibroblast growth factor (FGF-2). Aside from their role in helping neurons bind and promoting neurogenesis, they trigger a molecular chain reaction that produces endothelial cells, which make up the inner lining of blood vessels and thus are important for building new ones. These inroads expand the vascular network, bringing each area of the brain that much closer to a lifeline and creating redundant circulation routes that protect against future blockages. Second, exercise introduces more nitric oxide, a gas that widens the vessels' passageways to boost blood volume. Third, the increased blood flow during moderate to intense activity reduces hardening of the brain arteries. Finally, exercise can to some extent counteract vascular damage. Stroke victims and even Alzheimer's patients who participate in aerobic exercise improve their scores on cognitive tests. Starting when you're young is best, but it's never too late.

2. It regulates fuel. Researchers at the Karolinska Institute conducted a nine-year study of 1,173 people over age seventy-five. None of them had diabetes, but those with high glucose levels were 77 percent more likely to develop Alzheimer's.

As we age, insulin levels drop and glucose has a harder time getting into the cells to fuel them. Then glucose can skyrocket, which creates waste products in the cells—such as free radicals—and damages blood vessels, putting us at risk for stroke and Alzheimer's. When everything is balanced, insulin works against the buildup of amyloid plaque, but too much encourages the buildup, as well as inflammation, damaging surrounding neurons.

Exercise increases levels of insulin-like growth factor (IGF-1), which regulates insulin in the body and improves synaptic plasticity in the brain. By drawing down surplus fuel, exercise also bolsters our supply of BDNF, which is reduced by high glucose.

3. It reduces obesity. Aside from wreaking havoc on the cardiovascular and metabolic systems, body fat has its own nasty effects on the brain. The CDC estimates that 73 percent of Americans over sixty-five are overweight, and, given the potential problems obesity can lead to—from cardiovascular disease to diabetes—the agency is right in declaring it a pandemic. Simply being overweight doubles the chances of developing dementia, and if we factor in high blood pressure and high cholesterol—symptoms that often come along with obesity—the risk increases sixfold. When people retire, they figure they deserve a break after working their whole lives, and they start piling on the food. But what they don't realize is that having dessert with every meal is no treat. Exercise, naturally, counteracts obesity on two fronts: it burns calories, and it reduces appetite.

4. It elevates your stress threshold. Exercise combats the corrosive effects of too much cortisol, a product of chronic stress that can bring on depression and dementia. It also bolsters neurons against excess glucose, free radicals, and the excitatory neurotransmitter glutamate, all necessary, but they can damage cells if left unchecked. Waste accumulates and junks up the cellular machinery, and it starts turning out dangerous products—damaged proteins and broken fragments of DNA that trigger the latent and ultimately inevitable process of cell death that defines aging. Exercise makes proteins that fix the damage and delay the process.

5. It lifts your mood. More neurotransmitters, neurotrophins, and connectivity shore up the hippocampus against the atrophy associated with depression and anxiety. And a number of studies have shown that keeping our mood up reduces our chances of developing dementia. The evidence applies not only to clinical depression but also to general attitude. Staying mobile also allows us to stay involved, keep up with people, and make new friends; social connections are important in elevating and sustaining mood.

6. It boosts the immune system. Stress and age depress the immune response, and exercise strengthens it directly in two important ways. First, even moderate activity levels rally the immune system's antibodies and lymphocytes, which you probably know as T cells. Antibodies attack bacterial and viral infections, and having more T cells makes the body more alert to the development of conditions such as cancer. Population studies bear this out: The most consistent risk factor for cancer is lack of activity. Those who are physically active, for instance, have a 50 percent lower chance of developing colon cancer.

Second, part of the immune system's job is to activate cells that fix damaged tissue. When it's out of whack, these damaged spots fester, and you're left with chronic inflammation. This is why, if you're over fifty, your blood will be tested for C-reactive proteins as part of your standard physical. These proteins are a sign of chronic inflammation, a primary risk factor for cardiovascular disease and Alzheimer's. Exercise brings the immune system back into equilibrium so it can stop inflammation and combat disease.

7. It fortifies your bones. Osteoporosis doesn't have much to do with the brain, but it's important to mention because you need a strong carriage to continue exercising as you age, and it is a largely preventable disease.

Osteoporosis afflicts twenty million women and two million men in this country. More women every year die from hip fractures — a vulnerability of osteoporosis — than from breast cancer. Women reach peak bone mass at around thirty, and after that they lose about 1 percent a year until menopause, when the pace doubles. The result is that by age sixty, about 30 percent of a woman's bone mass has disappeared. Unless, that is, she takes calcium and vitamin D (which comes free with ten minutes of morning sun a day) and does some form of exercise or strength training to stress the bones. Walking doesn't quite do the job — save that for later in life. But as a young adult, weight training or any sport that involves running or jumping will counteract the natural loss. The degree to which you can prevent the loss is impressive: one study found that women can double their leg strength in just a few months of weight training. Even women in their nineties can improve their strength and prevent this heartbreaking disease.

8. It boosts motivation. The road to successful aging really begins with desire, because without the desire to stay engaged and active and alive, people quickly fall into the death trap of being sedentary and solitary. One of the problems of getting older is the lack of challenges, but with exercise we can continually improve and push ourselves.

Exercise counteracts the natural decline of dopamine, the key neurotransmitter in the motivation and motor systems. When you move, you're inherently boosting motivation by strengthening the connections between dopamine neurons, while at the same time guarding against Parkinson's. This really underscores the idea that if you're not busy living, your body will be busy dying. It's important to have plans and goals and appointments, and this is why sports such as golf and tennis are great. They require constant self-monitoring and the motivation to improve.

9. It fosters neuroplasticity. The best way to guard against neurodegenerative diseases is to build a strong brain. Aerobic exercise accomplishes this by strengthening connections between your brain cells, creating more synapses to expand the web of connections, and spurring newly born stem cells to divide and become functional neurons in the hippocampus. Moving the body keeps the brain growing by elevating the supply of neurotrophic factors necessary for neuroplasticity and neurogenesis, which would otherwise naturally diminish with age. Contracting your muscles releases factors such as VEGF, FGF-2, and IGF-1 that make their way from the body into the brain and aid in the process. All these structural changes improve your brain's ability to learn and remember, execute higher thought processes, and manage your emotions. The more robust the connections, the better prepared your brain will be to handle any damage it might experience.

LISTEN TO MY MOTHER

My family got its first television when I was about eight, but we never camped out in front of it. We weren't allowed to; my mom would say, "Don't just sit there; go out and play." We ate fish every week, not only because we were Catholic, but also because even back then it was known as "brain food." And the nuns at school drilled in the importance of staying mentally active with the mantra "An idle mind is the devil's workshop." Long before the benefit of scientific proof, the stern women of my upbringing fixated on the three pillars of a healthy lifestyle: diet, exercise, and staying mentally active. In that sense, the prescription for living a long and rich life hasn't changed much. But now we know so much more about the why and how that it's hard to ignore the advice.

DIET: EAT LIGHT, EAT RIGHT

The one proven way to live longer is to consume fewer calories—at least if you're a lab rat. In experiments in which rodents eat 30 percent fewer calories, they live up to 40 percent longer than animals allowed to eat as much as they want. "Our control group is really overfed and underexercised," says neuroscientist Mark Mattson, pointing out that group is "a good match for a lot of the American population." A study in monkeys that began eighteen years ago in the experimental gerontology lab at the National Institute on Aging suggests the same holds true in primates. And one human trial showed that asthma patients on a restricted diet for two months—three meals one day and only five hundred calories the next—had fewer markers of oxidative stress and inflammation in their blood (and their asthma symptoms improved). This finding supports the theory that imposing mild stress on the cells—in this case depriving them of fuel—makes them more resilient to future challenges and reduces free radicals. "It's kind of like exercising for an hour every day," Mattson says. "It's a mild stress, but as long as there's a recovery period, it's good."

He is cautious about telling people to skip meals, but that's what he does: no breakfast, a salad for lunch, and a normal dinner, for a total of about two thousand calories. It's likely that people of normal weight wouldn't benefit as much, and anyone over fifty should be careful about malnourishment because they're losing muscle and bone anyway. But if you're overweight, you're inflicting damage to your brain.

As for what you do eat, there are certain foods that activate cellular repair mechanisms, as I mentioned in chapter 3. Cumin, garlic, onions, and broccoli, for instance, all contain toxins meant to keep pests at bay, but the levels are low enough that they trigger a beneficial stress response. The same holds true for free radical–fighting foods such as blueberries, pomegranates, spinach, and beets — it's the toxins as well as the antioxidants that ultimately lead to cellular repair. Green tea and red wine are beneficial in the same way.

The rest of your plate should be balanced with whole grains, proteins, and dietary fats. Low-carb diets may help you lose weight, but they're not good for your brain. Whole grains have complex carbohydrates that supply a steady flow of energy rather than the spike and crash of simple sugars, and they're necessary to transport amino acids such as tryptophan into the brain. As you learned in chapter 4, tryptophan is a precursor necessary for the production of serotonin, and it and other important amino acids come from protein.

The brain is made up of more than 50 percent fat, so fats are important too, as long as they're the right kind. Trans fat, animal fat, and hydrogenated oils gum up the works, but the omega-3s found in fish are enormously beneficial. Population studies have shown that countries in which people eat a lot of fish have lower incidence of bipolar disorder. And some people use omega-3s as a stand-alone treatment for mood disorders and ADHD. One study showed that people who eat fish once a week slow the yearly rate of cognitive decline by 10 percent. The Framingham Heart Study followed nine hundred people for nine years and found that those who ate three meals with fish oil per week were half as likely to develop dementia. Omega-3s lower blood pressure, cholesterol, and neuronal inflammation, and they elevate the immune response and BDNF levels. You can find omega-3s in deep-water fish such as salmon, cod, and tuna, or you can take a daily supplement that contains 1200 mg of eicosapentaenoic acid (EPA) and 200 mg of docosahexaenoic acid (DHA) — the two key omegas.

Vitamin D is being recognized not only for its importance in strengthening bones but also as a measure against cancer and Parkinson's. I would recommend 1000 IU (international units) of vitamin D and for women, 1500 mg of calcium. I would also recommend taking vitamin B with at least 800 mg of folate, which improves memory and processing speed.

PHYSICAL EXERCISE: STEADY DOES IT

For anyone over sixty, I recommend exercising almost every day. In retirement, why not? Six days a week would be ideal, but make it fun rather than work. It's a good idea to use a heart rate monitor. They're invaluable for keeping track of your progress, and this is both motivating and reassuring. You're not left wondering whether you've done enough, at the proper intensity. They come with instructions, but essentially you subtract your age from 220 to find your theoretical maximum heart rate, and then use that number to figure out how hard you should be working. (I'll explain more about using them in the next chapter.)

Your overall strategy should include four areas: aerobic capacity, strength, and balance and flexibility. You should consult with a doctor or trainer who knows your history, but I can give you some solid guidelines.

AEROBIC: Exercise four days a week, varying from thirty minutes to an hour, at 60 to 65 percent of your maximum heart rate. At this level, you'll be burning fat in the body and generating the ingredients necessary for all the structural changes in the brain I've discussed. Walking should be perfectly adequate, but do it outside with a friend if possible. Whatever you choose, try to find something that you will enjoy over the long haul. Try a more intense pace for two days a week — 70 to 75 percent of your maximum — for twenty to thirty minutes. If you haven't been exercising, you'll want to build up to this speed, and that's fine. Consistency is probably more important than intensity. "You don't have to work as hard as you may think you do," suggests Kramer. "If you can work harder and run rather than walk, that's great. But if you can't, walking is what we've looked at, and it can have some fairly dramatic effects."

STRENGTH. Hit the weights or resistance machines twice a week, doing three sets of your exercises at weights that allows you to do ten to fifteen repetitions in each set. This is critical for preventing and counteracting osteoporosis: even if you do all the aerobic training in the world, your muscles and bones will still atrophy with age. A Tufts University study of women fifty to seventy years old showed that those who participated in strength training for a year added 1 percent of bone density in their hips and spine, while the sedentary group lost 2.5 percent of the density in those areas. If you don't have experience with resistance training, it's a great idea to get a trainer for the first month or to find some type of instruction — good form is important in avoiding injuries. Activities that involve bouncing or jumping also help strengthen your bones: tennis, dancing, aerobics class, jumping rope, basketball, and, of course, running.

BALANCE AND FLEXIBILITY. Focus on these abilities twice a week for thirty minutes or so. Yoga, Pilates, tai chi, martial arts, and dance all involve these skills, which are important to staying agile. Without balance and flexibility, your ability to stick with an aerobic and strength-training regimen will diminish. In lieu of an activity, you can do drills on an exercise ball, balance board, or Bosu, which is half a rubber ball that you stand on to challenge your core muscles. Remember Harold, the octogenarian skier? He trained on the Bosu for his most recent ski trip.

MENTAL EXERCISE: KEEP LEARNING

My advice here is to keep challenging your mind. You know by now that exercise prepares your neurons to connect, while mental stimulation allows your brain to capitalize on that readiness. It's no coincidence that study after study shows that the more education you have, the more likely you are to hang onto your cognitive abilities and stave off dementia. But it's not necessarily about the diploma. It's just that those who have spent a lot of time in school are more likely to remain interested in learning. Tucked within those statistics are plenty of people who didn't go to college yet nurture intense interests in the world around them. The most inspiring evidence of this comes from an urban health study called Experience Corps conducted by epidemiologists at Johns Hopkins University. They enlisted 128 predominantly African American women between sixty and eighty-six with low education levels and socioeconomic status and trained them to teach grade school children reading skills, library skills, and so on. Not only did the children improve their scores on standardized tests, but the health of the volunteers substantially increased: half of the women who were using canes stopped; 44 percent reported feeling stronger; the amount of time they spent watching television dropped 4 percent; and they reported a significant increase in the number of people they could turn to for help.

Volunteering is beneficial because it involves social contact, which is inherently challenging for the brain. Anything that keeps you in contact with other people helps you live better and longer — statistics show a tight inverse relationship between sociability and mortality. Novel experiences demand more from your brain, and this builds its ability to compensate. You get more Miracle-Gro, more connections, more neurons, and more possibilities.

There was a nun named Sister Bernadette who died of a heart attack at age eighty-five in the mid-1990s. Along with more than six hundred other nuns, she donated her brain to science as part of an ongoing study conducted by epidemiologist David Snowdon, who memorialized the School Sisters of Notre Dame, in Mankato, Minnesota, in his inspiring book, *Aging with Grace*. The nuns constantly challenge their minds, with vocabulary quizzes, mental puzzles, and debates about public issues, and many of them live to be one hundred or more. The interesting thing about Sister Bernadette is that she scored in the ninetieth percentile on cognitive tests right up until she died, but when her brain was examined postmortem, it showed massive damage from Alzheimer's disease. Tissue from her hippocampus to her cortex was riddled with plaque and neurofibrillary tangles to the most extensive degree, and she also carried the ApoE4 gene variant. In other words, she should have been utterly lost to the ravages of dementia. Yet despite the damage in her brain, she remained mentally sharp.

Snowdon points to the notion of cognitive reserve as a possible explanation. This is the brain's ability to adapt and compensate for damage by recruiting other areas to help with tasks. By teaching until a late age and staying mentally active, Sister Bernadette almost certainly trained her brain to work around the genetic hand she was dealt. Her example, like that of my mother, is one to live by.

10
The Regimen
Build Your Brain

I'VE TRUMPETED THE astounding impact of aerobic activity on the brain in the hope that if you understand what's actually going on up there when you go for a run, you'll develop a genuine motivation to lace up your sneakers every day. Or you'll go for a swim or hop on your bike or do whatever you enjoy doing to work up a sweat. I want nothing less than to get you hooked.

The point I've tried to make — that exercise is the single most powerful tool you have to optimize your brain function — is based on evidence I've gathered from hundreds and hundreds of research papers, most of them published only within the past decade. Our understanding of how the brain works has absolutely exploded in this relatively short period, and it's been an incredibly exciting time for anyone interested in the human condition. For me, personally, researching this book has redoubled my enthusiasm for the benefits of exercise and replaced intuition with hard, scientific fact.

As an illustration of just how new this territory is, I'll go back to the story of neurogenesis, the once-heretical theory that the brain grows new nerve cells throughout life. "Ten years ago people weren't even convinced that it happened," says neurologist Scott Small. It was at his Columbia University lab, in 2007, where they witnessed telltale signs of neurogenesis for the first time in live humans. "Five years ago people said, OK, it might happen, but is it really meaningful? Now there isn't a week that goes by where there's not another study that shows neurogenesis has some kind of effect on the brain."

In his study, Small put a group of volunteers on a three-month exercise regimen and then took pictures of their brains. By manipulating a standard MRI machine's processing—essentially zooming in and cocking the shutter open—he captured images of the newly formed capillaries required for nascent neurons to survive. What he saw was that the capillary volume in the memory area of the hippocampus increased by 30 percent, a truly remarkable change. The real breakthrough here might turn out to be the ability to map neurogenesis without slicing into the brain, which could shift the research focus from lab rats to people. The new technology should allow scientists to test the impact of any given variable on neurogenesis, such as how much exercise is necessary. "Is it one hour a week? Is it every morning? Is it only a grueling marathon regimen that maximizes neurogenesis?" Small asks. "We simply don't know that. No one knows that. Now by having this tool that can measure neurogenesis indirectly, we can actually try to optimize an exercise regimen."

Several years down the line, that is. At this point, he and his peers view exercise primarily as a surefire trigger to increase the growth of new cells. It's something they use to observe another process; most haven't gotten around to studying exercise per se.

It's a similar story with many of the other positive effects of exercise I've discussed, from increasing neurotransmitters and neurotrophic factors to releasing factors from the muscles that build new capillaries in the brain and encourage synaptic plasticity. William Greenough, the neuroscientist who saw that exercise caused neurons to sprout new branches with an electron microscope in the early 1970s, will tell you there's no question that aerobic exercise is great for your brain. And he's pretty confident that it's important to include complex motor movements (such as aerobic dance or martial arts) in your routine. But he can't give specific recommendations just yet.

And that's OK. We don't need to rely exclusively on neuroscientists to get started. First of all, there are certain conclusions we can draw from the work they've already done. Second, there is instructive evidence from other fields. Research from kinesiologists to epidemiologists shows again and again that the better your fitness level, the better your brain works. Charles Hillman proved that fit children score better than unfit children on cognitive tests of executive function; Arthur Kramer showed that getting in shape increases brain volume of older adults; and population studies including tens of thousands of people of every age show that higher fitness levels relate directly to positive mood and lower levels of anxiety and stress.

When people ask me how much exercise they should do for their brain, I tell them the best advice is to get fit and then continue challenging themselves. The prescription for how to do that will vary from person to person, but the research consistently shows that the more fit you are, the more resilient your brain becomes and the better it functions both cognitively and psychologically. If you get your body in shape, your mind will follow.

Does that mean you have to look like an underwear model to enjoy the brain benefits of exercise? Not at all. In fact, many of the most convincing studies use walking as the mode of exercise. But I focus on getting fit because we know with certainty that having a normal body mass index and a robust cardiovascular system optimizes your brain. Any level of activity will help, certainly, but from a practical standpoint, if you're going to bother doing something for your brain, you might as well do enough to protect your body against heart disease, diabetes, cancer, and the like. Body and brain are connected. Why not take care of both?

BORN TO RUN

In his book *Racing the Antelope: What Animals Can Teach Us about Running and Life,* biologist Bernd Heinrich describes the human species as an endurance predator. The genes that govern our bodies today evolved hundreds of thousands of years ago, when we were in constant motion, either foraging for food or chasing antelope for hours and days across the plains. Heinrich describes how, even though antelope are among the fastest mammals, our ancestors were able to hunt them down by driving them to exhaustion—keeping on their tails until they had no energy left to escape. Antelope are sprinters, but their metabolism doesn't allow them to go and go and go. Ours does. And we have a fairly balanced distribution of fast-twitch and slow-twitch muscle fibers, so even after ranging miles over the landscape we retain the metabolic capacity to sprint in short bursts to make the kill.

Today, of course, there's no need to forage and hunt to survive. Yet our genes are coded for this activity, and our brains are meant to direct it. Take that activity away, and you're disrupting a delicate biological balance that has been fine-tuned over half a million years. Quite simply, we need to engage our endurance metabolism to keep our bodies and brains in optimum condition. The ancient rhythms of activity ingrained in our DNA translate roughly to the varied intensity of walking, jogging, running, and sprinting. In broad strokes, then, I think the best advice is to follow our ancestors' routine: walk or jog every day, run a couple of times a week, and then go for the kill every now and then by sprinting.

Your choices aren't limited to these modes of aerobic activity, naturally, but I think they're helpful categories to distinguish between low-intensity (walking), moderate-intensity (jogging), and high-intensity (running) exercise. If you want to make the most of your time and effort, you'll need a way to accurately judge your level of exertion along these divisions. When I talk about walking, or low-intensity exercise, I'm referring specifically to exercising at 55 to 65 percent of your maximum heart rate. By my definition, moderate intensity falls in the range of 65 to 75 percent, while high intensity is 75 to 90 percent. The upper end of high-intensity exercise is sometimes painful but always powerful territory that has gained a lot of scientific interest recently.

Unless you're content to toil away on an exercise machine with built-in pulse sensors, the only way to accurately gauge your level of intensity is to use a heart rate monitor. These gadgets are the cornerstone of the revolutionary physical education program in Naperville, and they're simple enough that elementary school kids there know how to use them. They consist of a chest-strap sensor that picks up your heartbeat, and a digital watch that receives the signal and displays the number of beats per minute on its screen. Let's say your regimen calls for a high-intensity run. If you're forty-five years old, your theoretical maximum heart rate would be about 175, based on the rough formula of 220 minus your age. If you calculate 75 percent and 90 percent of your maximum, the lower and upper limits for a high-intensity workout are 131 and 158. This is your target heart rate zone for the workout. All you have to do is punch in these limits on the watch, which is no more difficult than setting the time, and then adjust your pace according to what the monitor tells you. The watch will beep if your heart rate strays from the desired zone. It's a reasonably accurate way to listen to your body.

Heart rate monitors are cheap, easy to use, and indispensable for anyone who's seriously interested in tapping the potential of exercise. It's also just nice to know that you're doing enough and not too much. But again, the question is how much? Public health recommendations, from the Centers for Disease Control to the American College of Sports Medicine, suggest doing some form of moderate aerobic exercise for thirty minutes at least five days a week. But I think they're hedging their bets. Americans are so inactive that the experts are wary of providing guidelines that are too stiff, for fear that the whole country will give up. "Everyone wants to know the minimum amount they can do for the bang," says Duke University exercise physiologist Brian Duscha, who was inundated with media requests after he published a study showing that walking as little as three hours a week has cardiovascular benefits. "I try not to overwhelm people because they quit." He also points out to anyone who'll listen that increasing the duration or the intensity carries even greater fitness gains.

Duscha is an expert in cardiovascular health, but he says the same thing almost every neuroscientist cited in these pages has said: "A little is good, and more is better." The *best*, however, based on everything I've read and seen, would be to do some form of aerobic activity six days a week, for forty-five minutes to an hour. Four of those days should be on the longer side, at moderate intensity, and two on the shorter side, at high intensity. And while there's conflicting evidence about whether high-intensity activity, which can force your body into anaerobic metabolism, impacts thinking and mood, it clearly releases some of the important growth factors from the body that build up the brain. So, on the shorter, high-intensity days, include some form of strength or resistance training. These days should not be back to back; your body and brain need recovery time to grow after high-intensity days. In total, I'm talking about committing six hours a week to your brain. That works out to 5 percent of your waking hours.

That said, I do agree with experts like Duscha that the most important thing is to do *something*. And to start. This last bit may sound obvious, but for the sedentary—especially if inactivity is due to depression—taking that first step may seem impossible. For some people it's a catch-22: they can't start because they don't have the energy, and they don't have the energy because they're not exercising. I've seen this happen with some of my patients, and it's a very real problem, not simply an issue of willpower. The key is to attack the business of starting as a challenge in itself.

It's well established that it's easier to work out with someone else, whether you're running with a friend, cycling with a group, or walking with a neighbor. Beyond that, there are several new studies showing that the neurological benefits I have described are greater when exercising with someone else. I tell patients who are really stuck to consider getting a personal trainer for a while, because then they're less likely to cancel a session (you pay regardless of whether you show up as scheduled, and money is a great external motivator). Write exercise into your schedule just like a dentist appointment. After a while, your brain will absorb it into your routine, just like brushing your teeth.

If you haven't been active, I think the best way to begin is to start walking. Take the stairs instead of the elevator, park at the back of the lot, and go for a stroll around the block at lunchtime. There's a decades-old health initiative called ten thousand steps that encourages people to use a simple pedometer to calculate how much they walk every day, as a way to work exercise into their routine without having to think much about it. Based on the average stride of 2.5 feet, ten thousand steps is close to five miles. It's a clever way to begin to get in shape without even setting aside much extra time. And it works. Counting your steps, like weighing yourself or using a heart rate monitor to guide your efforts, helps keep you focused and motivated, especially if you understand what your body and brain are doing at various intensity levels.

WALKING

The process of getting fit is all about building up your aerobic base. The more you work your heart and lungs, the more efficient they become at delivering oxygen to your body and brain. With the increased blood flow, of course, comes the chemical cascades that produce serotonin, brain-derived neurotrophic factor (BDNF), and other nourishing molecules.

If you start walking an hour a day at between 55 and 65 percent of your maximum heart rate, you'll naturally increase the distance you walk in that time period and gradually get in shape. At this level, you're burning fat as fuel, and this begins to gear up your metabolism. When a body carries too much fat, its muscles build up resistance to insulin, which exacerbates the fat buildup and curtails the production of insulin-like growth factor (IGF-1). A study published in 2007 from the University of Michigan reported that a single session of aerobic activity completely reversed insulin resistance the next day. The researchers compared muscle biopsies from before and after the session, and they also saw that the exercised fibers produced proteins important for fat synthesis. They don't know how long the effect lasts, but the findings underscore how even a minor amount of activity sparks a positive domino effect.

When you place a demand on your body and your muscles sense the need for more fuel, all sorts of good things happen. Low-intensity, fat-burning exercise also pumps up free tryptophan in the bloodstream, which, you'll recall, is a necessary ingredient for the production of mood-stabilizing serotonin. This level of activity also changes the distribution of norepinephrine and dopamine. When you look at this in the evolutionary context of Heinrich's endurance predator, it makes elegant sense: while tracking their prey, our ancestors needed to have the patience, optimism, focus, and motivation to keep at it. All these traits are influenced by serotonin, dopamine, and norepinephrine.

Walking will make you feel more invested in the world around you. Before long, you'll want to get out there even more. One simple way doctors gauge the fitness of their patients is to see how far they can walk in six minutes. But researchers at the University of Alabama School of Medicine found that people improve so rapidly that the best way to get an accurate reading is to have them do two practice walks first. Which is to say, you might be pleasantly surprised at how quickly you begin to cover more ground.

Once you work up to an hour at a pace just above where you can carry on a conversation, you're ready to add in moderate-intensity exercise. When you challenge yourself at that level, you'll be able to do more not only while you're working out but also in every domain of your life. You'll have more vigor and energy, less negativity, and you'll develop a greater sense of control. Above all, if you're in an active mode, you won't be sitting at home, isolated and stagnant.

JOGGING

Once you move into moderate-intensity activity, between 65 and 75 percent of your maximum heart rate, your body shifts from burning fat alone to also burning glucose, and the muscle tissue develops microtears as a result of the stress. All the cells in your body and brain are in a constant state of damage and repair, but the metabolic demands at this level ratchet up the response. Your body knows it needs a more robust oxygen-delivery system, so the muscles release vascular endothelial growth factor (VEGF) and fibroblast growth factor (FGF-2), and they start cells dividing to make more tissue for more blood vessels—the new capillaries that Scott Small captured on film. In lab cultures, researchers have found that VEGF and FGF-2 activate cells to make more blood vessels within just two hours of exposure. In the brain, in addition to building new blood vessels, these two factors also encourage cell binding and neurogenesis.

Inside your brain cells, the higher activity level triggers the release of metabolic cleanup crews, producing proteins and enzymes that dispose of free radicals, broken bits of DNA, and inflammation factors that can cause the cells to rupture if left unchecked. More and more, research suggests that taking antioxidants in pill form may not be helpful—and may actually be harmful—but most people don't realize that aerobic exercise is a way of creating your own supply of antioxidants inside the cells themselves. And anti-oxidants are only part of the story. Assuming there's an adequate recovery period, the repair response to exercise leaves your neurons stronger.

Moderate-intensity exercise also releases adrenaline into the bloodstream. In the untrained person, the hypothalamus-pituitary-adrenal (HPA) axis is activated. This is the fight-or-flight stress response I described in chapter 3, in which your body is on high alert, and cortisol begins coursing through your brain. At moderate levels, cortisol cues up the cellular machinery of learning in the interest of recording a situation that your body assumes is important to survival. But if cortisol remains chronically elevated, it's toxic to nerve cells. BDNF is the best defense for your neurons. By elevating restorative chemicals with a moderate workout, you're strengthening the circuits in your brain and tuning up the HPA axis so that it isn't as trigger-happy to future incidents of stress. Likewise, the immune system becomes stronger and better prepared to handle a genuine assault on the body — fighting off everything from colds to cancer.

Another factor from the body that comes into play here is atrial natriuretic peptide (ANP). Produced by the muscles of the heart itself when it's really pumping, ANP travels through the bloodstream and into the brain, where it helps to further moderate the stress response and reduce noise in the brain. It's a potent part of a cascade of chemicals that relieve emotional stress and reduce anxiety. Along with pain-blunting endorphins and endocannabinoids, the increase in ANP helps explain why you feel relaxed and calm after a moderate aerobic workout. When you talk about burning off stress, these are the elements at work.

At this level, you're tearing things down and building them back up again, stronger than they were before. It's crucial to build in recovery time, so your body and brain have the opportunity to bounce back.

RUNNING

At high intensity, between 75 and 90 percent of your maximum heart rate, your body enters into a full-fledged state of emergency, and its response is appropriately powerful. It's also within this range, usually at the upper end, that metabolism shifts from aerobic to anaerobic, in which your muscles go into a state of hypoxia because they can't pull enough oxygen from the bloodstream. Since oxygen is necessary for the efficient burning of glycogen, without it your muscles begin to burn creatine and glycogen stored directly in the muscle tissue, a messy mechanism that creates a buildup of lactic acid (the burn you feel in your thighs and chest). The so-called anaerobic threshold occurs at different intensities for different people, but during a high-intensity workout, you want to be flirting with that thigh burn (after warming up) but staying just below it throughout your run. And while physiologists can't give a magic heart rate at which your body crosses over from aerobic to anaerobic, one recent study by kinesiologist Panteleimon Ekkekakis of Iowa State University showed that the most reliable marker of this metabolic change was when subjects reported that their level of exertion became "somewhat hard." It sounds vague, but Ekkekakis has found the correlation to be remarkably consistent. Another way to judge this level is to keep in mind that even though going at just below your anaerobic threshold is "somewhat hard," it shouldn't be so hard that you couldn't keep up the same pace for thirty minutes to an hour.

If you want to really challenge yourself by mixing in intervals, you'll sprint above that threshold for short bursts in the middle of your high-intensity session.

One of the key differences between moderate and high-intensity exercise is that once you get closer to your maximum, and especially when you get into the anaerobic range, the pituitary gland in your brain unleashes human growth hormone (HGH).

This is what life-extension groups call the fountain of youth. The levels of HGH naturally secreted into the bloodstream decrease over your life span, so that by middle age they dwindle to a tenth of what they were during childhood, for both women and men. And a sedentary lifestyle exacerbates this decline: high levels of cortisol, insulin resistance, and excess fatty acids in the bloodstream all clamp down further on the hormone's release.

HGH is the body's master craftsman, burning belly fat, layering on muscle fiber, and pumping up brain volume. Researchers believe it can reverse the loss of brain volume that naturally occurs as you age. Athletes such as Olympic sprinters and football players are essentially spiking their levels of HGH when they do interval training, doping the natural way. The result is a buildup of fast-twitch muscle fibers, which adds power to their movements. Also, recruiting new muscle fibers enhances your metabolism overall, and your body retains the improved ability to burn fat and carbohydrates after the interval training.

Normally HGH stays in the bloodstream only a few minutes, but a session of sprinting can keep the level elevated for up to four hours. In the brain, HGH balances neurotransmitter levels and boosts the production of all the growth factors I've mentioned. But it seems to have the most dramatic impact on IGF-1, the evolutionary linchpin tying together activity, fuel, and learning. It gets into the very cell nucleus and switches on genes that crank up the mechanisms of neuronal growth.

Psychologically, this is where you "confront the self," in the words of my colleague Robert Pyles, the psychiatrist and marathoner I wrote about in chapter 3. By going beyond where you thought you could, straining and stressing and lingering in that pain for even just a minute or two, you sometimes transcend into a rarefied state of mind, in which you feel like you can conquer any challenge. If you've ever experienced the phenomenon of runner's high, it probably came in response to a near maximal effort on your part. The euphoric feeling is likely due to the mixture of extremely high levels of endorphins, ANP, endocannabinoids, and neurotransmitters pumping through your system at this intensity. It's the brain's way of blocking everything else out so you can push through the pain and make the kill.

High-intensity exercise toughens you up, both physiologically and psychologically. It's the reason why we climb mountains and sign up for fitness boot camps and go on Outward Bound trips. Yet you don't need to go to such extremes to reap the rewards I'm talking about. One study from the University of Bath, in England, found that adding a single spurt of sprinting for thirty seconds — in this case while pedaling on a stationary bike — generated a sixfold increase in HGH, which peaked two hours after the sprint.

And a recent study by neurologists at the University of Muenster, in Germany, reported that interval training improves learning ability. During the course of a forty-minute treadmill run, volunteers did two three-minute sprints (separated by two minutes at a lower intensity). Compared with subjects who stayed at low intensity, the sprinters had significantly higher increases in BDNF as well as norepinephrine. Accordingly, in cognitive tests immediately following the run, the sprinters learned vocabulary words 20 percent faster. So even a small dose of pushing yourself to the limit has profound effects on your brain.

As good as all this sounds, interval training isn't something you can just jump off the couch and do. You really need to have a rock-solid aerobic base and talk to your doctor about what you're planning. It's not a good idea to put that kind of strain on your heart if it's not used to it. Depending on what sort of shape you're in, I'd say you should have at least six months of six-days-a-week aerobic activity before mixing in interval training. But again, your doctor needs to clear you for takeoff.

NONAEROBIC ACTIVITY

I haven't devoted much space to the discussion of nonaerobic exercise because, frankly, there is very little research into how it affects the brain in terms of learning, mood, anxiety, attention, and the other issues I've covered. It's difficult to get rats to pump iron or do yoga, so scientists are restricted to studying humans, which means they can't biopsy brain tissue after the experiments. They have to rely on blood samples and behavioral tests, which leave much more room for interpretation. The results we do have on nonaerobic exercise aren't as robust as those for aerobic exercise.

That said, strength training is obviously important for building muscles and protecting joints, and practices such as yoga and tai chi improve balance and flexibility—all of which contribute to your body's ability to remain active throughout life. A very recent study among older adults found that lifting weights twice a week for six months made participants stronger and actually reversed aspects of the aging process at the genetic level. The genes responsible for producing some of the key factors for brain growth (VEGF, FGF-2, and IGF-1) acted as if they were thirty years old, not sixty-five.

Most of the brain research on resistance training has focused not on learning and memory, but on mood and anxiety. In one study from Boston University dating back ten years, researchers put a group of older adults on a twelve-week strength-training program (three sessions a week) and measured various aspects of psychological and cognitive function. They concluded that in addition to improving muscle strength by about 40 percent, the regimen lowered anxiety and improved mood and confidence levels, but had no significant effect on thinking ability. Around the same time, another study from the Institute of Psychology at the University of Bern, in Switzerland, tested the effects of eight weeks of strength training. Doing a ten minute warm-up followed by eight weight-machine exercises once a week improved psychological well-being and had a slight effect on memory. And according to follow-up tests, the results remained for a year regardless of continued exercise levels. But there were too many variables for the researchers to conclude that strength training had a measurable effect on memory.

The intensity level of the strength training seems to affect the results, in that moderate weights have been shown to have a more positive impact than heavy weights, at least in a small group of older women. Other research has shown that high-intensity strength training actually increases anxiety levels in both men and women. In this case, high intensity was defined as lifting 85 percent of the maximum weight possible, but many studies haven't defined this crucial variable. A study published a few years ago in the *American Journal of Sports Medicine* showed that cross training—combining thirty minutes of weights with thirty minutes of stationary cycling—improved anxiety levels, but the study was designed in such a way that it was impossible to know what caused the change. And almost all of the studies on this topic have used populations of older adults, who are prone to pronounced improvements because their muscles are naturally diminished when they start.

One factor clearly affected by strength training is HGH. A recent study looked at the hormone levels during weight training versus aerobic activity in well-trained men. Doing squats doubled HGH levels compared with running at high intensity for thirty minutes, and I think this will turn out to have important implications for exercise recommendations.

There is even less research into the effect of rhythm, balance, and skill-based activites on the brain. Small studies have shown that yogic breathing reduces stress and anxiety levels, and tai chi reduces the activity of the sympathetic nervous system (judging from heart rate and blood pressure). One recent study used MRI scans on eight yoga practitioners and found that their levels of the neurotransmitter gamma-aminobutyric acid (GABA) increased 27 percent after a sixty-minute session. GABA is the target of

Xanax-like drugs and is very involved with anxiety, so this may be part of the reason why yoga helps some people relax. Much of the evidence from this realm is anecdotal, but I'm sure that as neuroscientists delve deeper into the brain they'll discover pathways that connect the dots.

STICKING WITH IT

Statistics show that about half of those who start up a new exercise routine drop out within six months to a year. Not surprisingly, one of the biggest reasons seems to be that people often jump in at high intensity, which makes them feel bad physically and emotionally, and then they simply quit. The kinesiologist Ekkekakis has focused much of his research on the relationship between exercise intensity and discomfort. People vary as to how they feel leading up to the shift from aerobic to anaerobic metabolism, but he has found that once they cross the line almost everyone reports negative feelings on psychological tests and high ratings on scales of perceived exertion. It's your brain putting you on alert that there's an emergency. The point is, if you feel lousy even at lower intensity levels, don't take on interval training in the early stages of your new routine. (Again, it's more important to do something than nothing.)

And don't get down on yourself if you don't love exercise — you may be genetically predisposed to dislike it. In 2006 European researchers compared physical activity levels of 13,670 pairs of identical twins and 23,375 pairs of fraternal twins, who share only half the same genes. They found that 62 percent of the variation in whether the twins tended to exercise was explained by different genes. Other research has found that gene variations have an impact on whether you enjoy the feeling of exercise, whether you stick with it once you begin, and even whether you notice a dramatic improvement in mood. Among the many genes involved, researchers have focused on one related to dopamine, the reward and motivation neurotransmitter, and another that controls BDNF expression. People with the dopamine variation may have reward deficiency syndrome, which robs them of that rush of pleasure they imagine everyone else at the gym is experiencing. And if your BDNF signaling is off, the mood-improving mechanism of exercise may be sluggish. I offer this information not as an excuse, but as a reminder — all of us can rewire our brains by taking action. It's not as easy as when we were kids, but it's clearly possible.

Exercise immediately increases levels of dopamine, and if you stay on some sort of schedule, the brain cells in your motivation center will sprout new dopamine receptors, giving you newfound initiative. You're wearing in new neural pathways or perhaps refurbishing ones that are rusty from disuse, and it only takes a few weeks to solidify a habit. Exercise can become a self-reinforcing behavior that helps you trump your genes. The truth is, your genes are only one part of a very complex equation, and you have control over many of the other variables.

It's a similar story with BDNF: it may take longer for you to get over the hump of forming a routine and to feel good while you're exercising, but once you do your brain becomes more and more efficient at producing Miracle-Gro. Carl Cotman, the neuroscientist who runs the aging center at the University of California, Irvine, discovered that the hippocampus has what he calls a "molecular memory" for producing BDNF. In a three-month trial, he measured BDNF levels of lab rats on various exercise routines, comparing daily versus alternate-day wheel running, and looking at the effect of quitting for several weeks. The impetus for the study was based on his wry observation that although most lab experiments use daily exercise, "In humans, exercise patterns are generally less rigorous, and rarely follow a daily consistency."

He came to a number of telling conclusions. First of all, daily exercise ramps up BDNF more rapidly than alternate days — at a rate of 150 percent versus 124 percent after two weeks. Curiously, after a month, the alternate-day exercisers had caught up with the daily group. When the rats stopped exercising, regardless of the routine, it took just two weeks for levels of BDNF to return to baseline. But the most interesting finding was that when the exercisers were allowed access to their running wheels again, BDNF levels shot back up in just two days (137 percent above normal for the daily exercisers, and 129 percent for the alternate-day group). This is what he means by molecular memory: if you've had the experience of exercising regularly, your hippocampus can get back up to speed very quickly.

Cotman concluded that every day is best but that even intermittent exercise works wonders. And I think it's important for people to recognize that exercise isn't an all-or-nothing proposition. If you miss a few days, or even a week or two, picture your hippocampus cranking out high levels of BDNF after just your second day back at it.

STRENGTH IN NUMBERS

One of the best ways to get on a roll is to get in a group. The stimulus of social interaction starts your neurons firing like nothing else — it's complicated, challenging, rewarding, and fun. And when you combine this sort of mental activity with the priming effect of exercise, you're maximizing the growth potential of your brain. Exercise cues up the building blocks of learning, and social interaction cements them in place.

Princeton neuroscientist Elizabeth Gould, a pioneer in the field of neurogenesis whose research focuses on how experience and environment change the brain, has studied the different effects of exercise on animals living alone versus those living in a group. She has found that social interaction has a powerful impact on neurogenesis. In one experiment, after twelve days of running, rodents housed in social groups showed a significant increase in neurogenesis over others that exercised just as much but were kept in isolation. In fact, the isolated runners had the same low level of cell proliferation as group-housed controls that didn't exercise. The reason has to do with the stress hormone cortisol. In her study, published in *Nature Neuroscience* in 2006, Gould found that while all the runners had elevated cortisol during exercise, levels for the isolated group were also high at other times of the day. In other words, cortisol won out over neurogenesis in the isolated condition, but social support "blunted the reactivity" of the HPA axis and kept the stress hormone from interfering with growth. Does that mean that going for a run by yourself is bad news? Not at all.

Remember that exercise itself is a stressor, which activates the HPA axis and may elevate cortisol. So is isolation. It seems that the cumulative stress of running *and* being alone elevated cortisol to such an extent that it prevented neurogenesis from happening— perhaps because the rats didn't have an adequate recovery period. Making matters worse, the rodents were sedentary to begin with, and going from no activity to running several kilometers a day presents a major new stress on the system.

When Gould extended the experiment beyond the initial twelve days, it was an entirely different story. She found that if the rats were kept in the same conditions, over the long term the isolated runners' systems caught up. Somewhere between twenty-four and forty-eight days of running, the rate of neurogenesis leveled out between the isolated and social groups. She speculates that one explanation might have to do with serotonin, which is increased by social interaction and in turn enhances neurogenesis. Both isolation and prolonged exposure to cortisol reduce the number of serotonin receptors in the hippocampus. It might be the case that even though running increases levels of serotonin, without enough receptors for it to plug into the neurons, it can't do its job.

Gould is attempting to elucidate extremely complex relationships between stress, environment, and exercise, and there are several crucial points to take away from her study: First, it's important to start slow if you haven't been exercising and have a lot of other stress in your life. Second, social support has a powerful effect on the brain and can both prevent the negative impacts of stress and clear the way for exercise to ramp up the machinery of growth, so keep connected to keep your connections. And third, if you keep at your routine, your system will adjust itself to take advantage of the activity.

Of course, Gould emphasizes that there are limitations to drawing conclusions from animal studies. "Rodents are really different from humans," she says. "If you give a rat or a mouse access to a running wheel, every single one of them will run. That's not true of humans. A lot of them buy treadmills that become coatracks."

While it's true that we're born to run, we're also programmed to take advantage of bountiful periods and to conserve our energy for the long stretches of foraging and hunting that surely lurk around corner. It's not that the instinct to plop down on the couch suddenly appeared in our DNA in the past hundred years; it's that our modern environment is incongruous with our genes. Food is never far from hand—foraging requires ten steps to the fridge, not ten miles across the savannah—so it's important to replace the need to work for it with the demands of aerobic exercise.

But don't become a lab rat. Reserve your treadmill running for rainy days or for times when you can't arrange to do something with other living, breathing human beings. Joining a team or setting a goal of doing a 10K charity run and training with a group of friends adds a sense of obligation, which can be a powerful motivator. Back in Naperville, Zientarski teaches cooperation, not competition, but for some adults being part of a team can get them hooked, whether it's a three-on-three city basketball program, an adult soccer league, or Masters swimming.

Maybe walking with the one you love will turn out to be the key, or maybe you've always wanted to learn tae kwon do, or maybe, like Naperville Central graduate Jessie Wolfrum, you'll discover a passion for the infinite challenge of rock climbing (which requires a partner). She was lucky, as a high school student, to have eighteen different activities to choose from. You, on the other hand, are lucky to be able to choose from any form of activity you can imagine. The beauty of exercise is that the more you do, the more you'll imagine yourself doing.

STAY FLEXIBLE

Naturally, it's important to stretch, but it's just as important to keep your mind flexible. The problem with any routine, of course, is that it goes against nature. The world around us is constantly changing, and it's difficult to keep doing the same thing over and over. I wouldn't ask you to. The best strategy is to do something almost every day, yet stay flexible within that framework, allowing it to bend, so it doesn't break. By mixing up your program and trying new activities, you'll continue to adapt and challenge yourself. My own experience with exercise is a perfect example of what can go wrong and also what can go right.

I grew up in western Pennsylvania, in the era during which that area produced football stars such as Joe Namath, Mike Ditka, and Tony Dorsett. I played the major sports—football, basketball, and baseball—but was more of a hardworking scrub than a game maker. I found my athletic calling on the tennis courts and played constantly with my best friend and doubles partner all through high school. I was supposed to play at Colgate, but I broke my arm and leg in a car accident just before heading off to college. My arm required two surgeries that kept me out of commission for several years. I quit competing in tennis and didn't do much else athletic for at least a decade.

I got back into being active during my residency, which was during the running craze surrounding Bill Rogers's success and the popularity of the Boston Marathon. Running made me hungry to play tennis again, and then I got into squash with a couple of colleagues, including my good friend and longtime collaborator Ned Hallowell. We played three times a week for almost twenty-five years, competing, cajoling, and encouraging one other. We were all extremely busy, but our squash appointments were sacrosanct. It was a magical stretch.

About seven years ago, I tore the rotator cuff of my right arm beyond repair, and I can no longer swing a racket. I started lifting weights for rehabilitation, and it was the first time that I consistently went to the gym. I started going three or four times a week, doing forty minutes or so on the Stairmaster or elliptical trainer, and continuing with the weights two days a week. Then I kicked it up a notch and blocked out an hour a day, but I missed the camaraderie of our squash games. Ned coerced me into signing on with his personal trainer, Simon Zaltzman. He is a relic: a former boxing coach with a thick Russian accent and what seems to be an endless imagination for challenging me.

I got into the habit of doing weights plus crunches and balance exercises two times a week, three times when I'm really focused. The other days I do forty minutes on the elliptical trainer, or on the treadmill when I want to add in some intervals.

While researching this book, I learned about the magic of HGH and how sprinting might get me where I really wanted to be. Two days a week, I started including a handful of sprints during my treadmill runs, and let me tell you, they hurt. Just writing about it makes me cringe a little, but it has been well worth the extra effort. After one month of this business, I lost the final ten pounds I'd been after for years—it peeled right off my midsection. Not that I was overweight. It's just that nothing I tried seemed to budge my spare tire. Now, on two of my aerobic days (and no more than two) I do just twenty minutes of jogging, interspersed with five sprints of twenty to thirty seconds each in which I run as fast as I can. When people want to know what to do if they're short on time, I tell them this story.

Although I'm almost sixty, I feel much younger, and if I could get Art Kramer to scan my brain, I'm sure it would look younger too. I'm doing everything I can to keep my prefrontal cortex, and everything it's connected to, pumped up. I miss days, certainly, but I try not to miss two in a row. When I just can't make it to the gym, my wife and I take the dogs for a thirty-minute speed march instead of a ten-minute walk. For Jack and Sam, our spunky and inexhaustible Jack Russells, my dalliance is their great good fortune. Little do they know.

Afterword
Fanning the Flames

IN WRITING THIS book, I remain hopeful for the future of our country and for our kids. The spark begins with them, naturally, and I have seen what happens when it catches fire, as it has in Naperville. Nineteen thousand kids, and only 3 percent of them overweight. And they're smarter for it. Despite the fact that we are in danger of eating ourselves to death and killing our brains in the process, things are changing. Physical activity and exercise are making their way back into American life. In 2007, as his first mission in office, the new governor of Florida, Charlie Crist, pushed through legislation to get grade schoolers exercising at least thirty minutes a day. He teamed up with Shaquille O'Neal to spread the word. The superintendent of schools in Kansas City, Missouri, decided to establish daily physical education throughout the district after seeing how it reduced violence and improved test scores nearly overnight at one of his inner-city elementary schools. In other parts of the country, legislators are holding hearings on stemming the tide of inactivity in an effort to beat the grim statistics.

Even in medicine, we're beginning to see people officially take exercise seriously. In his inaugural address in 2007, the president of the American Medical Association, Ronald M. Davis, urged all AMA members to read a pamphlet entitled "Exercise Is Medicine," so they can help each and every patient plan an exercise regimen. In psychiatry, it's happening too. In the May 2007 issue of the *Journal of Clinical Psychiatry,* the association offered, for the first time, a Continuing Medical Education course related to physical activity: Exercise for Mood and Anxiety Disorders. CME courses are a key venue for doctors to keep up on the latest medical science, so there should be plenty to discuss in this new one. Every day there are more studies testing exercise as an intervention for mental health issues. Exercise physiologists are being added to staff at assisted-living centers and nursing homes, and there aren't enough personal trainers to meet demand at health clubs.

We're taking our cues from neuroscience and our inspiration from all the people who have already seen what exercise can do for their brains. My hope is that everything I've laid out in these pages will encourage you to grab your gym bag instead of the remote, or spend time on the field rather than on the sidelines. From your genes to your emotions, your body and brain are dying to embrace the physical life. You are built to move. When you do, you'll be on fire.

Acknowledgments

I am grateful to the many people who had a hand in helping me produce *Spark*.

Phil Lawler and Paul Zientarski, along with their staff at the schools in Naperville District 203, were a constant source of inspiration. The fitness-oriented model of physical education they created brings life to the growing evidence of how exercise builds a better brain. Their revolutionary slant on gym class has improved the academic performance and the lives of their students. Equally important, it has established an atmosphere in which PE teachers have the freedom and desire to try new ideas that might have a positive impact on their charges. In spirit and in practice, Lawler and Zientarski are researchers themselves—relentless, courageous, and curious.

A special mention goes to Neil Duncan and his students from Zero Hour PE, as well as their literacy teachers Maxyne Kozil and Debbie St. Vincent, for opening a window onto the Naperville ethos.

Anne Flannery and the staff at PE4life were a great help in pointing out other communities where the Naperville model has taken root. Titusville, Pennsylvania, is just one example, and Tim McCord spent hours relating the effects of the program there.

Hearty thanks go to all the neuroscientists and experts we interviewed, for their generosity of time and assistance. They include James Blumenthal, Alyson Bond, Craig Broeder, Darla Castelli, Eero

Castrén, Maureen Dezell, Rodney Dishman, Wayne Drevets, Andrea Dunn, Brian Duscha, Panteleimon Ekkekakis, Fred Gage, Sam Goldstein, Elizabeth Gould, William Greenough, Thom Hartmann, Charles Hillman, Marian Joels, Dean Karnazes, Arthur Kramer, Helen Mayberg, Bruce McEwen, Ina Mullis, Peter Provet, Robert Pyles, Amelia Russo-Neustadt, Terry Robinson, Jennifer Shaw, Tracey Shors, Scott Small, June Smedley, Dean Solden, John Tavolacci, Gene-Jack Wang, Jennifer Weuve, and Martin Wojtowicz. I am particularly thankful to Carl Cotman and Mark Mattson, whose research and insights opened my eyes to the intracellular benefits of exercise. I can only hope to emulate their passion for discovery.

I would like to express my deep gratitude to the patients and friends who agreed to share their stories, and to the many others whose experiences were incorporated in short form to make these pages come alive.

My editor, Tracy Behar, and the staff at Little, Brown were so excited about this book that they volunteered for a six-month study on how increasing their levels of exercise would affect them. Their unbridled interest added fuel to the fire, which helped bring the manuscript across the finish line. Brooke Stetson, my conduit to Tracy, was ever-enthusiastic in her encouragement.

My agent and tireless advocate, Jill Kneerim, brought shape to my vision and helped to structure the book from the outset. And Elisabeth Weed connected Eric Hagerman and me in the first place.

Thanks to those who read parts of the manuscript and provided valuable feedback, including my sister, Veronica Crain; Dave Goodrich; Allen Ivey; Eric's mother, Judy Sinderson; and his friend Stephen Milioti, among many others. Jacob Sattelmair took valuable time away from his PhD program to discuss the science on many occasions. I am indebted to Ned Hallowell, first my student and then my teacher, who demanded many years ago that we hold sacred a weekly slot for squash and fellowship. It was Ned too, who introduced me to Simon Zaltzman, a master trainer who never fails to challenge me with a torture du jour. Thanks to Ben Lopez for his friendship, his thoughts on the book, as well as his generosity in turning over his house by the sea when I was looking for a place to start.

My assistant, Mary Haroun, has been invaluable — guiding me through the myriad details associated with the process. She handled a multitude of issues that freed up the time I needed to devote to the book. Most important, she has been a friend, a rock, and a source of strength, cheering me forward at every turn. Likewise, I couldn't have done without the friendship, support, and emergency technical assistance of Mary's husband, Majdi Haroun.

Eric proved to be more than a collaborator and cowriter. He challenged my hop-and-skip brain to stay focused and pushed me to make him understand what I was trying to communicate. His ability to grind it out and to help crystallize two gigabytes' worth of scientific papers was essential in bringing this book to heel. We spent many a day together hammering out the pages and growing in the process. His writing has been superb and his passion indispensable. I appreciate the sacrifice in time spent away from his beloved partner, Christelle.

One of the hardest parts of writing this book was sustaining the energy over a two-year period, something that would have been insurmountable without the support, encouragement, and love of my family, friends, and colleagues. Thank you all for being there when I needed you. I am especially grateful to my wonderful daughters, Jessica and Kathryn, and new son-in-law Aaron Cohen, for their suggestions on the manuscript and unwavering support.

And finally, to my wife, Nancy, who while finishing her own book somehow found the strength and understanding to allow me the time to finish mine. She was my champion throughout.

Glossary

adrenal glands. Small organs located just above the kidneys. One part of the glands produces and releases epinephrine (adrenaline) to initiate the stress response; another part produces and releases cortisol and cortisol-like hormones when directed by signals from the HPA axis to ramp up the stress response. See also **HPA axis.**

aerobic metabolism. The long-term mode of energy conversion defined by the sufficient availability of oxygen to burn fuel — first fat, and then fat and stored glucose — to feed active muscle cells. It occurs at low to moderately intense levels of physical activity and can be sustained over long periods of time.

anaerobic metabolism. A mode of energy conversion defined by lack of sufficient oxygen to convert fat and glucose into usable body fuel. When the body is pushed so fast and so hard that the working muscles' demand for fresh oxygen outstrips the circulating blood's ability to deliver it, the muscles begin to burn fuel inefficiently.

anandamide. A neurotransmitter in the body and the brain that binds to cannabinoid receptors. Tetrahydrocannabinol (THC), the active ingredient in marijuana, also activates these receptors. When the cannabinoid receptors are activated, they help the body and the brain manage pain, mood, and pleasure.

atrial natriuretic peptide (ANP). A naturally occurring hormone that is produced in the heart and the brain. When the heart rate increases, ANP production increases, and ANP is released into the bloodstream. It passes through the blood-brain barrier to counteract the action of certain elements of the stress response. It blunts stress and anxiety and helps to regulate mood. See also **blood-brain barrier.**

blood-brain barrier. A web of capillaries with tightly packed cells that prevents the easy transmission of some nutrients and substances from the bloodstream to the brain. It filters out toxins and infections.

brain-derived neurotrophic factor (BDNF). A protein produced inside nerve cells when they are active. It serves as Miracle-Gro for the brain, fertilizing brain cells to keep them functioning and growing, as well as spurring the growth of new neurons.

cerebellum. A small, densely packed part of the brain that contains half of the total nerve cells in the entire brain and is involved in integrating sensory and automatic motor function. It is constantly busy, updating and computing incoming and outgoing information. Within the past two decades, scientists have recognized that the cerebellum is involved in maintaining rhythm and continuity for many brain functions, such as emotions, memory, language, and social interactions, as well as allowing us to walk in a straight line. I call it the rhythm and blues center.

cortex. The brain's thin outer layer of gray matter, just six cells thick. As the last portion of the human brain to have evolved, it is the seat of rapid computing and guides the rest of the brain. Neurons throughout the brain extend their axons to connect with the cortex and thus inform it about a wide range of mental activity.

cortisol. The primary long-acting stress hormone that helps to mobilize fuel, cue attention and memory, and prepare the body and brain to battle challenges to equilibrium. Cortisol oversees the stockpiling of fuel, in the form of fat, for future stresses. Its action is crucial to our survival. At high or unrelenting concentrations, cortisol has a toxic effect on neurons, eroding the connections between them and breaking down muscles and nerve cells to provide an immediate fuel source.

dopamine. A neurotransmitter that is vital to movement, attention, cognition, motivation and pleasure, and addiction.

endocannabinoids. A class of hormones known as the brain's own marijuana. They are similar to endorphins in that they blunt pain, although they are metabolized much more quickly than tetrahydrocannabinol (THC) and thus have relatively fleeting effects.

endorphins. Hormones produced in the body and the brain that serve as natural morphine. They are released when the body and the brain are taxed, to block pain signals so we can push through physically uncomfortable situations. They affect many physiological functions, such as pleasure, satisfaction, and bliss.

epinephrine. Also called adrenaline, it is both a neurotransmitter in the brain and a hormone released from the adrenal glands. It is immediately released during the stress response, to prepare the nervous system to react to challenges to survival.

fibroblast growth factor (FGF-2). A protein that is produced and released in the body and the brain when tissues are stressed. Like vascular endothelial growth factor (VEGF), FGF-2 helps create more blood vessels and other tissues. FGF-2 is involved in initiating the process of stem-cell division that is necessary for neurogenesis, and it also encourages long-term potentiation (LTP) and the formation of memories. See also **long-term potentiation, neurogenesis,** and **vascular endothelial growth factor.**

gamma aminobutyric acid (GABA). The principal inhibitory neurotransmitter of the brain. It inhibits overactivity of all nerve cells, particularly those in the limbic system, which is home to the amygdala, the emotional center. Many antianxiety medications target GABA receptors. GABA is involved in anxiety, aggression, mood, and seizure control.

glutamate. The principal excitatory neurotransmitter of the brain. It is critical to cell binding and, thus, neuroplasticity.

hippocampus. The way station for many aspects of learning and memory. It gathers incoming stimuli from throughout the brain, cross-references the new information with stored information, and bundles it together as a memory that is then sent to the prefrontal cortex for processing. In recent years, it has been shown to be a critical component in the biology of stress and mood, since it contains a vast number of cortisol receptors and is the first step in regulating the feedback loop of the fight-or-flight response. Its close relationship with cortisol leaves the hippocampus particularly vulnerable to the ravages of stress and aging. Conversely, it is one of only two structures in the brain that clearly produces its own new nerve cells. See also **neurogenesis.**

HPA axis. A signaling route from the hypothalamus to the pituitary gland to the adrenal gland, which controls the stress response. It is important to such vital functions as fuel regulation and the immune system. See also **adrenal glands, hypothalamus,** and **pituitary gland.**

human growth hormone (HGH). A hormone known as the master of all hormones, it is vital to the growth and development of all cells in the brain and the body into adulthood and is intimately involved with building the body. It controls fuel allocation and counteracts the natural cellular atrophy of aging.

hypothalamus. A small gland situated above the pituitary gland that makes and secretes hormones to signal the pituitary to release hormones and other factors. It is a switching station that translates instructions from the brain, carried by neurochemical signals, into hormonal signals that travel through the bloodstream and direct biological imperatives such as sex, hunger, sleep, and aggression. See also **pituitary gland.**

insulin-like growth factor 1 (IGF-1). A hormone that is produced mainly in the liver and works closely with human growth hormone (HGH) and insulin to stimulate cell growth and counteract natural cell deterioration.

long-term potentiation (LTP). The cellular mechanism for learning and memory that requires the strengthening of brain cells' ability, or potential, to send a signal across the synaptic gap. It is crucial for the process of cell binding and, thus, communication. See also **synapse.**

Maximum heart rate. The physiological limit of the number of times an individual's heart can beat in one minute. This number is useful in correctly calculating the intensity of physical exertion. It can be determined in a physiology lab, by working to exhaustion. More commonly, recreational athletes use a theoretical maximum derived by subtracting their age from 220.

mitochondria. Tiny structures within every cell nucleus that function as the cell's furnace, using oxygen to convert glucose into a usable fuel during aerobic metabolism. When oxygen levels are insufficient, fuel conversion is shifted outside the mitochondria for anaerobic metabolism, a far less efficient process than aerobic metabolism. See also **aerobic metabolism** and **anaerobic metabolism.**

neurogenesis. The process of stem cells dividing and developing into functional new brain cells, or neurons, in the brain. That this happens in the adult human was firmly established in 1998, and it is believed to be confined to part of the hippocampus and another brain area called the subventricular zone, which is associated with the sense of smell. See also **stem cells.**

norepinephrine. A neurotransmitter that affects arousal, alertness, attention, and mood. Norepinephrine signals activate the sympathetic nervous system and sharpen the senses. See also **sympathetic nervous system.**

pituitary gland. A pea-size endocrine gland located just beneath the hypothalamus. It secretes hormones and factors that control other hormones throughout the body. See also **hypothalamus.**

prefrontal cortex. The region of the cortex located at the very front of the brain. As the last portion of gray matter to have evolved, the prefrontal cortex oversees the qualities that make us most human. It is the CEO of most brain functions, including, but not limited to, planning, sequencing, rehearsing, evaluating, and understanding. It is also the home of working memory, the brain's RAM, which is crucial to decision making. See also **cortex.**

serotonin. A neurotransmitter vital to mood, anxiousness, impulsivity, learning, and self-esteem. Often called the policeman of the brain, serotonin helps quell an overactive or out-of-control response in a wide range of brain systems.

stem cells. Undifferentiated cells that can develop into fully functioning new cells. In the adult human brain, they are located in a part of the hippocampus called the dentate gyrus, and in another area called the subventricular zone. Stem cells are encouraged to divide and develop into new neurons by fibroblast growth factor (FGF-2) and vascular endothelial growth factor (VEGF). See also **fibroblast growth factor, hippocampus,** and **vascular endothelial growth factor.**

sympathetic nervous system. A vast network of nerve cells that connect the brain to the body and are activated by norepinephrine. It is a component of the always-on autonomic nervous system, but its activity is dramatically increased during the stress response.

synapse. The junction of the axon and dendrite of two adjacent neurons. In the axon, electrical impulses are translated into chemical messengers—neurotransmitters—to carry instructions across the synaptic gap. At the dendrite, the neurotransmitter's signal is converted back into an electrical impulse, which prompts the receiving neuron to carry out a task.

vascular endothelial growth factor (VEGF). An important signaling protein produced and released in the body when tissues are taxed and there is not enough blood flow to fuel the demand. Like fibroblast growth factor (FGF-2), VEGF acts as a mitogen, signaling other cells to start dividing to make more blood vessels. Recently, scientists have learned that VEGF is also produced in the brain and is involved in cementing memories. See also **fibroblast growth factor.**

VO2 max. Maximal rate of oxygen consumption, a measure of the lungs' capacity to process oxygen; also known as aerobic capacity. It is the primary indicator of cardiovascular fitness.

Index

Index

Index

dendrites
    addiction and, 172
    aging and, 220, 222, 223, 225
    BDNF and, 132
    depression and, 129
    learning and, 36
    stress and, 67, 72, 74, 75–76
depression, 5, 22, 80, 102, 113–140, 164, 187, 196, 228, 235, 250
    addiction and, 182, 186
    ADHD and, 155, 161, 164
    aging and, 227, 229
    anxiety and, 88, 108, 109
    hormonal changes and, 214, 215
    learning and, 68
    medication and, 86–87, 101, 113, 115, 116, 119–121, 123, 124–127, 130–132, 134–135, 164
    menopause and, 211
    Parkinson's disease and, 231
    postpartum, 194, 203–208
    pregnancy and, 199
    serotonin and, 37
    stress and, 59, 76, 77, 82, 84, 235
    treatment for, 97
    types of, 115
de Vries, Herbert, 106
DHA (docosahexaenoic acid), 240
diabetes, 4, 76, 81, 198, 200, 209, 210, 218, 219, 220, 234, 247
Diagnostic and Statistical Manual of Mental Disorders (DSM), 125, 145–146, 193
diet, 73, 238–240
Dietrich, Arne, 182
Dishman, Rodney, 159, 165
disseminated sarcoidosis, 80–82
dopamine
    addiction and, 170, 171, 172, 176–177, 178, 179, 182, 183, 190
    ADHD and, 149, 150, 151, 158, 159, 160, 165–166
    aging and, 223, 228
    depression and, 115, 120, 121–122, 129, 135
    exercise and, 5, 136, 196, 237, 252
    genes and, 260–261
    hormones and, 194
    learning and, 38
    Parkinson's disease and, 151, 230, 231
    reward center and, 142
    stress and, 64–65, 79

Driven to Distraction (Ratey/Hallowell), 143–144
DSM (*Diagnostic and Statistical Manual of Mental Disorders*), 125, 145–146, 193
D2R2 allele, 176
Duke University, 7, 122, 164, 196, 228, 250
Duman, Ronald, 132
Duncan, Neil, 9–11
Dunn, Andrea, 137, 138
Duscha, Brian, 250
dyslexia, dyspraxia, and attention treatment (DDAT), 152–153

ECT (electroconvulsive therapy), 132–133
Edinburgh Postnatal Depression Scale (EPDS), 207
EEG (electroencephalogram), 25–26
Effexor (venlafaxine), 120
Ekkekakis, Panteleimon, 255, 260
electroconvulsive therapy. *See* ECT
Elliott, Frank, 154
El-Mallak, Olfat, 31
Embry-Riddle Aeronautical University (Daytona Beach, Florida), 26, 28
Emory University, 133
endocannabinoids, 182–183, 184, 257
End of Stress as We Know It, The (McEwen), 63, 77
endorphins, 64, 117–118, 121, 182, 183, 184, 200, 257
environmental enrichment, 46, 47, 49
EPA (eicosapentaenoic acid), 240
epinephrine, 63, 64, 65, 98, 103
Eriksson, Peter, 49
escitalopram (Lexapro), 121, 204
estrogen, 194, 197, 198, 206–207, 209, 212, 213, 227
Etnier, Jennifer, 212
excitotoxic stress, 72
executive function, 26, 53, 55, 142, 150, 151, 160, 213, 247
Exercise for Mood and Anxiety Disorders (CME course), 270
"Exercise Is Medicine," 269
exercise regimen, 245–251, 260–267. *See also* jogging; nonaerobic activity; running; walking
Exercising through Your Pregnancy (Clapp), 201
Experience Corps, 242

Index

Index

Index

Index

About the Author

John J. Ratey, MD, is an associate clinical professor of psychiatry at Harvard Medical School and has a private practice in Cambridge, Massachusetts. For more than a decade, he taught residents and Harvard medical students at Massachusetts Mental Health Center, where he served as the assistant director of resident training. He continues to teach psychiatrists as a regular instructor in Harvard's Continuing Medical Education program.

As a clinical researcher, he has published more than sixty papers in peer-review journals in the fields of psychiatry and psychopharmacology. In 1986 he founded the Boston Center for the Study of Autism, and in 1988 he formed a new American Psychiatric Association study group on aggression, which grew out of his research on novel drug treatments for aggressive behavior. During this time, Dr. Ratey lectured internationally on aggression and disturbances in the brain that affect social functioning.

Dr. Ratey and Dr. Edward Hallowell began studying ADHD in the 1980s and coauthored *Driven to Distraction: Recognizing and Coping with Attention Deficit Disorder from Childhood through Adulthood* (1994), the first in a series of books that demystify the disorder. Dr. Ratey coauthored *Shadow Syndromes* (1997) with Catherine Johnson, PhD, which explores the phenomenon of milder forms of clinical disorders. He also authored the bestselling book, *A User's Guide to the Brain: Perception, Attention, and the Four Theaters of the Brain* (2001), in which he explains how neuroscience affects emotions, behavior, and overall psychology.

Each year since 1998, Dr. Ratey has been selected by his peers as one of the best doctors in America. Most recently, Dr. Ratey received the 2006 Excellence in Advocacy award from the nonprofit group PE4Life, for his work to promote the adoption of regular, aerobic-based physical education.